Alimentary system	Eyes	Ears	Flexure	Torsion	Amnion	Allantois	Tail bud and limb buds
oregut 0.15 mm. long.							
oregut 0.3 to 0.4 mm. long.							
oregut 0.5 to 0.8 mm. long.	*Primary optic vesicles form.*						
oregut about 1.0 mm. long.	*Optic stalks begin to constrict at the base of the primary vesicles.*		*The head bends ventrally and sinks into the yolk.*				
oregut is about 1.3 mm. long.	*Differentiated into optic stalk and optic vesicle.*	*The auditory placodes begin to form from thickened ectoderm.*	*This cranial flexure increases in the region of the mid-brain.*	*The first signs of torsion appear in the head region.*	*The head amniotic fold begins to rise up and grow back.*		
he foregut is 1.5 mm. long nd there are indications of e first pair of visceral clefts.	*The ectoderm outside the primary vesicles thickens and becomes the rudiment of the lens.*	*The auditory placodes invaginate to form auditory pits.*	*Further sinking of the head into the yolk is prevented by the head twisting (torsion). Cranial flexure approaches 90°.*	*The head turns on to the left side. This torsion may reach the first two or three somites.*	*The head amniotic fold has grown back over the fore-brain.*		*The remains of the primitive streak begin contributing material posteriorly to the tail bud.*
he first pair of visceral clefts re distinct and the second air begin to form.	*The lens rudiments invaginate to form lens vesicles. The optic vesicles invaginate to form optic cups*	*The mouth of each auditory pit begins to constrict and auditory vesicles form.*	*Cranial flexure, i.e. angle between fore- and hind-brain, is 90°. Cervical flexure begins in hind-brain region and trunk flexure can also be seen.*	*The head is fully turned to the left. The first five to seven somites also exhibit torsion.*	*The hind brain and first few somites are covered by the amnion. Tail folds may begin to develop.*		*There is a distinct tail bud.*
he first and second visceral efts are clearly visible; the ird pair begin to develop. he hind gut appears.	*The mouth of the lens vesicle begins to close.*	*The mouth of the auditory vesicle is reduced to a small aperture.*	*Cranial flexure causes the fore-brain to be directed backwards close to the heart. Cervical flexure becomes a broad curve.*	*Torsion is apparent in somites eight to ten.*	*The sero-amniotic connection is somewhat attenuated. The amnion covers somites six to thirteen. Tail fold appears.*		*The tail bud begins to develop posterior to the hind gut.*
e first, second and third sceral clefts are present. e liver bud appears as do e tail gut and anal plate.	*The lens becomes cut off from the ectoderm. The optic cups are almost closed. The retina distinct. The eye is still anterior to the ear.*	*The auditory vesicle is connected to the small ectodermal aperture by the ductus endolymphaticus.*	*Cranial flexure is at its maximum. Cervical flexure increases. Trunk flexure is noticeable in the region of somites ten to twelve.*	*Torsion extends to somites eleven, twelve and thirteen or even further.*	*The head fold grows back and may lie anywhere between somites ten and eighteen. The tail fold begins to grow forward.*		*The tail bud can be seen projecting behind the hind gut. The limb buds appear as low swellings.*
e fourth pair of visceral efts develop. The liver bulge now conspicuous. Tail gut tends further into tail. oaca begins to form.	*The optic cup closes. The eyes now lie posterior to the ears.*	*The aperture closes.*	*Cranial flexure remains unchanged. Cervical flexure is about 100°. Trunk flexure develops into a broad curve. Caudal flexure begins.*	*Torsion as far back as somites fifteen to nineteen.*	*The head fold has extended to the region between somites sixteen to twenty-four. The tail fold has grown forward over somites 29–30.*	*Begins as an outgrowth of the hind gut in the cloacal region.*	*The tail bud begins to curve forward. The fore limb bud lies between somites 17–19 and the hind limb bud between somites 26–30.*
ur pairs of visceral clefts. e tail gut begins to degenere. The anterior and posterior estinal portals approach ch other, leaving an open estinal umbilicus of 3 mm. ng buds develop.	*The eyes, due to flexure, lie well posterior to the ears.*	*The auditory vesicle is pear-shaped with a narrow ductus endolymphaticus.*	*Caudal flexure causes tail to be at an angle of 90° to the body.*	*The whole posterior region exhibits some degree of torsion.*	*The head and tail folds meet or leave a small oval aperture over somites 26–28.*	*Allantoic stalk and vesicle. The vesicle enlarges after 72 hours.*	*Limb buds are now quite conspicuous and begin to exhibit nipple-shaped apices The hind limb bud extends to somite 32.*

An Atlas of Embryology

By the same authors

An Atlas of Histology

An Atlas of Invertebrate Structure

An Atlas of Plant Structure: Volume 1

An Atlas of Plant Structure: Volume 2

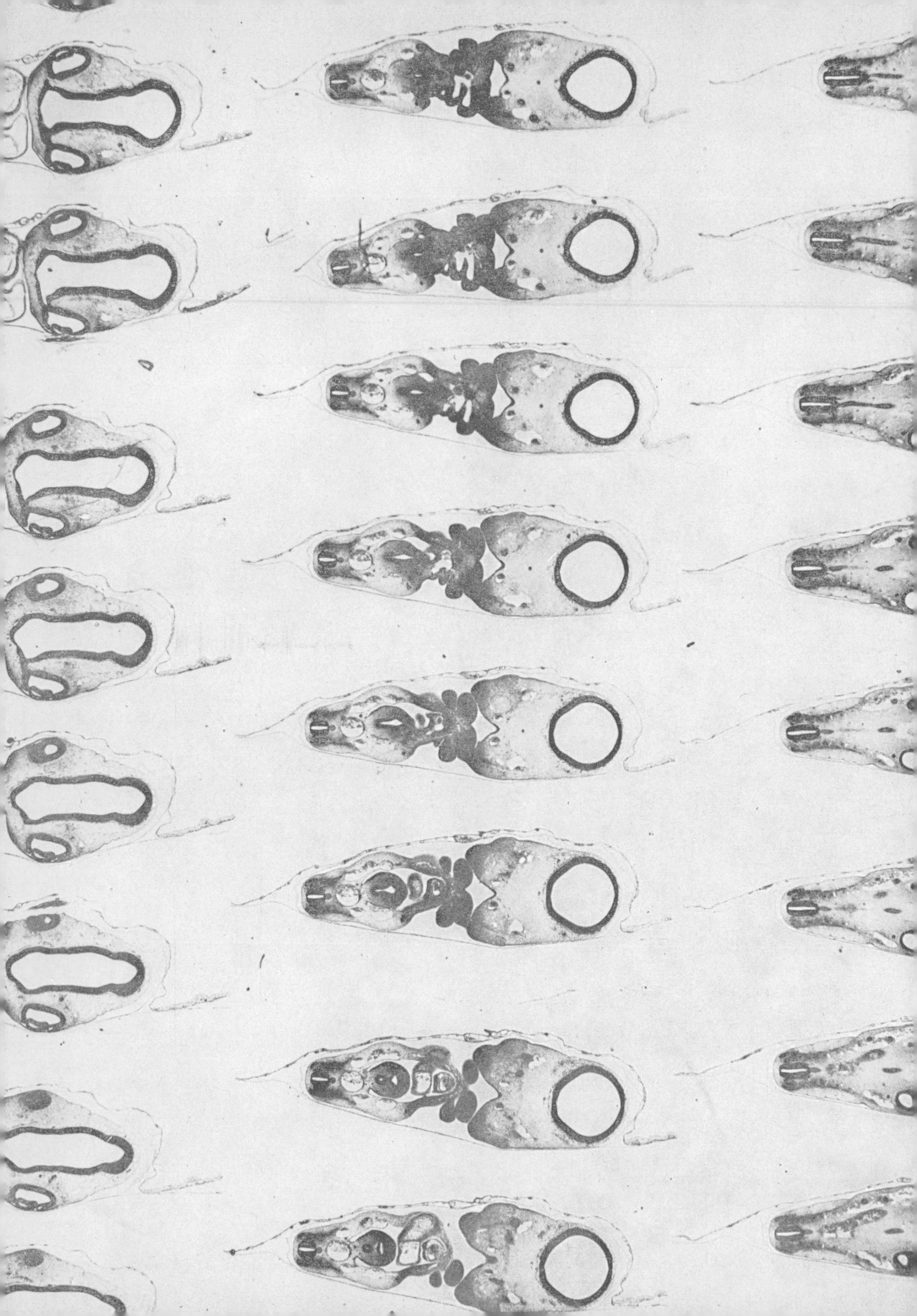

An Atlas of Embryology

by

W. H. Freeman, B.Sc., F.I. Biol.,

Head of the Biology Department, Chislehurst and Sidcup Grammar School
Chief Examiner, 'A' Level Zoology, London

and

Brian Bracegirdle, B.Sc., F.R.P.S., M.I. Biol.,

Principal Lecturer in Science, The College of All Saints, London N.17
Assistant Examiner, 'A' Level Zoology, London

H·E·B

Heinemann · London

Heinemann Educational Books Ltd
LONDON EDINBURGH MELBOURNE AUCKLAND TORONTO
SINGAPORE HONG KONG KUALA LUMPUR
NAIROBI IBADAN JOHANNESBURG
LUSAKA NEW DELHI

ISBN 0 435 60310 8

First published 1963
Reprinted 1964, 1965
Second Edition 1967
Reprinted 1969, 1970, 1972, 1975

The frontispiece shows
serial sections of 72 hour chick

Published by Heinemann Educational Books Ltd
48 Charles Street, London W1X 8AH
Printed in Great Britain by
Butler & Tanner Ltd, Frome and London

Preface to the first edition

This book consists of photomicrographs of sectioned and entire embryos of frog and chick, with large detailed drawings to correspond.

Descriptive embryology is still recognised as a necessary and valuable part of courses in zoology and biology leading to the General Certificate of Education at Advanced Level, and to first degrees. As teachers and examiners we have become aware of the difficulties experienced by students in interpreting the embryological structures seen under the microscope. The present book is intended to help overcome these difficulties, while at the same time summarising the descriptive embryology of frog and chick in sufficient detail for degree level. Care has been taken to label fully, and to make the drawings and photographs large enough for clearness.

It has become apparent that the embryology slides in general use are not of very high quality. For this reason, little attempt was made to obtain slides of better quality, but to use those normally confronting the student – in this way we hope to have improved the chances of artifacts being recognised as such. A large number of slides was looked through, but in the end we confined ourselves to a relatively small number of the more typical specimens. By doing this we were able to produce a book inexpensive enough for wide general use.

Each slide was photographed through the microscope, with special attention being paid to securing a flat field and good depth of focus – especially difficult with these rather large specimens. Not all the slides selected for inclusion were of a quality desirable for photomicrography, as will be obvious from the photomicrographs themselves; but we feel that this need be no great drawback, since students are often required to interpret these poorer-quality slides.

Each drawing was made completely independently of the photograph, directly from the slide. An accurate outline was obtained by microprojection, with the emphasis on line work, as it should be in students' drawings. Where it made for greater clarity, the drawing was diagrammatised, as in the case of some of the embryonic membranes. Later the drawing was compared with the photograph, and dotting was added where it seemed desirable for greater clarity. It will be seen that more detail appears in many of the drawings than in the corresponding photographs. This detail is obtainable only by the proper use of the fine focusing of the microscope at increased magnification, and should serve as a salutary reminder to the student that it is necessary for *him* to do the same to interpret *his* slides!

Much care and effort has been expended on the labelling of the drawings, and all the usual texts have been consulted. Even so, it was often necessary to have recourse to serial sections, where these were available. In many cases, none were, and so some errors are likely to remain, even though we were fortunate to have the fullest co-operation of Dr Ruth Bellairs, of University College, London, in checking the work. We are most grateful to Dr Bellairs for her great help; any errors remaining are, of course, the entire responsibility of the authors.

It would not have been possible to have produced this work from the slides already in our possession. For their kindness in making available extra material, we are deeply indebted to the following: Mr Charles Biddolph, Mr C. V. Brewer, Dr Ben Dawes, Mrs J. Froud, Mr George Gardener, Mr A. T. Green, Mr C. Heather, Dr Brian Lofts, Mr C. T. Pugsley, Mr A. R. Tindall, Mr H. Whate, and the Zoology Department of Wye College. To

Mr George Gardener we owe an additional debt for his early criticism and encouragement. We were likewise fortunate in our lettering artist, Mr Alan Plummer, who co-operated in a most wholehearted manner; and also in our Publishers – in Mr Alan Hill and Mr Hamish MacGibbon we found a most sympathetic support and facilitation of our aims. Last, but very definitely not least, we must thank our wives very sincerely indeed for their help and encouragement, and for their stoicism when surrounded for weeks on end by all the impedimenta of drawing and photomicrography.

W.H.F.
B.B.

September 1962

Preface to the second edition

In the four years which have elapsed since this book was first published, it has been necessary to alter very little of its contents, and we believe that our prime aim has been achieved – that is, to make more easy and more meaningful the interpretation of slides of embryological material in the laboratory.

For this new edition, we have added material on the development of amphioxus. This material is very scarce, and even though the type is the basic one in chordate embryology, most students never see any actual specimens of it. From time to time though, slides do become available; when they do they are often difficult to interpret because the material tends to be poorly fixed and stained, as will be apparent from the specimens we have had to use. Nonetheless, the inclusion of these stages has enabled us to introduce a short text in which the basic principles of embryological development are outlined as they apply to the chordates.

Some of the amphioxus slides are available from Philip Harris Ltd as colour transparencies for projection.

September 1967

W.H.F.
B.B.

Contents

A reference table of chick development is printed on the endpapers

Notes on basic embryology

These short notes outline basic concepts of chordate embryology, especially as exemplified in the specimens illustrated later in the book.

An egg is a living organism needing food both for energy and for growth, but since it is a single cell lacking the special feeding organs of the adult, its nutritional requirements have imposed many modifications on development. The primary source of food is yolk, stored in the egg while it is in the ovary; where this is insufficient for the completion of development a larval stage is quickly reached which hatches and becomes self-feeding. The later larva of amphioxus is an example. Water is available for aquatic larvae, but a group which has become terrestrial faces a very different situation. A compromise may be reached by returning to the water for egg-laying and larval growth, as in frog, but once full terrestrial life is attained water also must be stored in the egg. In reptiles and birds this is done by enclosing the egg within a shell. The egg membranes are permeable to oxygen and carbon dioxide, but relatively water-tight. Such a *cleidoic egg* is a closed system.

The chances of survival of the offspring are greatly increased by retaining the fertilised eggs inside the mother's body. In *ovoviviparity*, complete eggs are simply retained, being nourished by food stored in the egg itself. A further step is for the embryo in addition to absorb substances from the oviducal fluids – "uterine milk". In those dogfish where this occurs, the young are much heavier at birth than those where it does not. In *true viviparity* a direct connection forms between mother and embryo by means of a placenta. The huge increase in weight of a eutherian mammal and the advanced state of maturity at birth show how efficient this is as a means of supplying food and water to the embryo.

Kinds of Eggs.

Besides the normal cell-components, eggs usually have yolky food-reserves present, and the development of the chordate egg depends on the amount. *Isolecithal* eggs have little yolk, evenly distributed, as in amphioxus and most mammals. *Telolecithal* eggs with *total cleavage* have a moderate amount accumulated in one hemisphere, as in amphibian eggs. Telolecithal eggs with *meroblastic cleavage* have an enormous amount of yolk, the protoplasm being merely a germinal disc floating on its surface, as in reptiles and especially birds. These are arbitrary groupings, and there exists a continuous series between these stages. Where the distribution of yolk is uneven the protoplasmic half is called the animal hemisphere, the yolky half is the vegetal hemisphere. An imaginary line, the polar axis, can be drawn between the centre of each hemisphere; it is the earliest sign of symmetry in the egg, and it is probably a gradient of metabolic activity.

Fertilisation.

In the amniotes fertilisation is internal, being assured by special copulatory organs. in the lower groups fertilisation is external, relying on simultaneous liberation of eggs and sperms in large numbers and close proximity, aided by chemical attraction. At fertilisation, an initial activation of the egg occurs when sperms attach themselves to its surface, and one effects entry. Polyspermy is prevented by an alteration in the membranes, making them resistant to subsequent penetration. This initial activation is followed by syngamy, the nuclei of egg and sperm uniting to produce the diploid zygote nucleus, with equal contributions from each parent to the genetic constitution of the new individual. The act of fertilisation stimulates the egg to start developing.

Cleavage.

The first stage in development is a series of cell divisions which results in the production of many smaller cells from the zygote. During this process of cleavage the overall size of the embryo is unchanged. Throughout the animal kingdom, cleavage follows two main patterns, In the *Schizocoela* (polyclads, annelids, molluscs) cleavage is spiral, the spindle axes being oblique to the polar axis. With spiral cleavage goes determinate development (where the fate of each blastomere can be recognised very early in development), and unequal holoblastic cleavage. In the *Enterocoela* (echinoderms and chordates) cleavage is radial, where the spindle axes are in line with the polar axis or at right angles to it. With radial cleavage goes indeterminate development (where presumptive areas are recognisable only later in development). During cleavage nuclei divide mitotically with the long axis of each spindle lying along the long axis of cytoplasm. This axis is determined by the amount and distribution of the yolk in each cell. Cleavage furrows always appear at right angles to the spindle axis, and so although the inert yolk takes no part in cell division, it determines the planes of the cleavage furrows. *Holoblastic equal cleavage* occurs when in an isolecithal egg the entire egg divides to produce a number of similarly-sized cells, as in amphioxus. *Holoblastic unequal cleavage* occurs in telolecithal eggs having a moderate amount of yolk. Because of the yolky vegetal hemisphere, the early cleavage planes are shifted towards the animal pole, and are also delayed in the vegetal hemisphere. Even though the entire egg does cleave, these two factors make the cells at the animal pole smaller and more numerous than those at the vegetal. This cleavage is seen clearly in frog. *Meroblastic cleavage* occurs in eggs having very large amounts of yolk, where only the germinal disc cleaves, to give a small disc of cells on the surface of yolk, as in chick.

The Blastula.

In the early cleavage stages, the blastomeres are packed into a ball called the *morula* but later cells withdraw from the centre, to give the blastula containing a liquid-filled space called the blastocoel. In AMPHIOXUS this is large and central, the wall is one cell thick only, and the cells are about the same size. In FROG the blastocoel is smaller and less central while the wall is at least two cells thick, and the cells of the vegetal hemisphere are considerably larger than those of the animal. In CHICK the blastocoel is represented only by a slit between the germinal disc and the yolk; the blastodisc is several cells thick and slowly grows over the surface of the yolk to form the blastoderm.

Presumptive Areas.

It has been possible experimentally to plot fate maps of the blastula, which show what will eventually develop from any particular area. For example, the amphibian blastula has three main zones – a large one round the animal pole which will give the nervous system and epidermis; a large one round the vegetal pole which will form midgut and hindgut; and an equatorial marginal zone which will produce notochord, segmental muscles, mesodermal lining of body cavity, and foregut. The position of an area destined for a particular fate has nothing in common with the position of the organ in the adult.

Gastrulation and Morphogenetic Movements.

In gastrulation parts of the early embryo rearrange until they reach the places they will occupy in the adult, and the three germ layers of ectoderm, endoderm, and mesoderm are derived. During gastrulation morphogenetic movements occur; the rate of mitosis slows; growth slows; and the nucleus of each cell begins to assert control over the cell's activities. In AMPHIOXUS, the vegetal hemisphere invaginates until its wall reaches the animal hemisphere, giving a two-cell layer – ectoderm outside, and endoderm and prospective mesoderm within; eventually the blastocoel is eliminated. In FROG, gastrulation cannot occur like this, since the inert yolk prevents such complete invagination. Active cells from the marginal zone, at the dorsal lip of the blastopore, *roll* inside with an action quite unlike that in amphioxus. The lip gradually extends down over the vegetal region, until the blastopore has the form of a ring – the yolk-plug stage. This ring contracts in size ventrally until all the vegetal region is enclosed. In CHICK, the organ-forming part of the blastoderm is only a fraction of the blastodisc, the area pellucida, thus invagination is limited to this area, in a central *primitive streak*. Material in the upper layer, the *epiblast*, moves backwards and inwards towards the midline, and at the same time cells in the lower layer, the *hypoblast*, move forwards. When the streak has been formed single cells move down, giving a *total* effect of invagination, but in a process which is quite unlike either frog or amphioxus. The movement is called *immigration*, the whole primitive streak becoming a mass of moving cells, so that although the streak persists, it is being constantly replaced. Later, cells leave the streak faster than they are arriving, causing it to shrink backwards, until by the end of gastrulation it has almost disappeared. The movements of cells causing the above kinds of gastrulation are called *morphogenetic movements*, since they are the sum of irreversible movements by individual cells. Any one cell may change shape, may exert pressure on or adhere to its neighbours, may actively move, and may fail to adhere to certain other cells; all of these factors together causing groups of cells to exhibit selective affinities.

Germ Layers.

In AMPHIOXUS, as the gastrula elongates, part of the inner layer grows out to form paired mesodermal pouches, to give the classic three-layered condition. Later the cavities of the pouches become continuous to give a tube along each side of the embryo. In FROG the mesoderm as a whole splits off laterally first, and later grows to the midline. Dorsally, endoderm splits off from the hypoblast, to leave a layer of mesoderm, while ventrally lateral sheets grow down between yolk and ectoderm. At the anterior, mesoderm occurs not as a layer, but as loosely-arranged mesenchyme. Again, therefore, three layers are formed. In CHICK, when cells move through the primitive streak during gastrulation, they go to specific destinations, one of which is a layer of cells pushing out each side between epiblast and hypoblast. This layer becomes mesoderm, the epiblast becomes ectoderm, and the hypoblast becomes endoderm. This is a very rapid way of establishing mesoderm, and is an adaptation to the need to cover the yolk quickly with mesoderm, so that it may give rise to blood-vessels needed to carry yolky food to the embryo; the primitive streak is a short-cut.

By different means, therefore, correlated with the food requirements of the developing embryo, the three basic germ-layers characteristic of all triploblastic animals are laid down. *The student should note that triploblastic animals have three layers in the embryonic condition only, and then not for very long; and also that ectoderm, endoderm, and mesoderm are terms which are applicable only to these layers in the embryonic condition.*

Organisers and Organogeny.

In normal development *fate maps* show which part of an embryo will develop into which part of the adult – this is called the *prospective significance* of the part concerned. Experimentally though, it can be shown that a part removed from its normal site has the ability to develop into something quite different if grafted into a different place – this ability to develop in more than one way is called its *prospective potency*. The earlier in development the transplant is made, the greater the prospective potency of the part; conversely, as development proceeds, so a part is less and less able to match a site to which it is transferred. Eventually the prospective potency becomes identical with the prospective significance; this process is called *determination*, and it occurs for different tissues at different stages of development. Determination of a group of cells is caused by *embryonic induction* by adjacent tissues. This does not mean that the tissue being determined is passive, but on the contrary it must be in a particular state of reactivity called *competence*, and competence varies with time. Basically, a *primary organiser* is present, acting as a centre of induction: in most vertebrates the dorsal lip of the blastopore and the roof of the archenteron have this function. Induction *gradients* are found, since two substances are acting – an active one and a neutralizing one, their relative concentrations at any particular site deciding the effect. In this way, by a primary organiser gradient and later secondary gradients, cells are induced to differentiate in particular ways, and the organs of the adult body are laid down in the embryo.

The Fate of the Germ Layers.

The three basic layers undergo morphogenetic movements to give rise to the final organs, and these movements differ in type depending on whether they are made by epithelial cells (which are closely adherent), or by mesenchymal cells (which are much more individually mobile). *Epithelial cells form organs:* – (a) by local thickening of the layer, as in neural plate formation; (b) by separation of layers to leave a space, as in coelom formation; (c) by folding of a layer, as in early neural tube formation; (d) by fusion of once separate masses, as in later neural tube formation; and (e) by breaking up of layers to produce mesenchyme, as in formation of gut-wall and of neural crest. *Mesenchyme forms organs:* – (a) by aggregation into a mass, as in bone formation; (b) by attachment to another mass, as in the formation of skeletal capsules; and (c) by secondary rearrangement into epithelia, as in the formation of blood-vessels. As the result of all these movements, the ECTODERM GIVES RISE TO neural tube, from which develop brain, spinal cord, and nerves. Also from this layer develop the sensory parts of all sense organs, and the lens of the eye; the epidermis together with its derivates such as nails and hair and glands; and the lining membranes of nose, mouth, and anus THE ENDODERM GIVES RISE TO the epithelial lining (and only this *lining*) to the intestine and its outgrowths, including gills, larynx, lungs, thyroid, liver, pancreas, and bladder.

THE MESODERM is a layer with a complicated history. After formation, it splits into an outer *somatic* and an inner *splanchnic* layer, the space between being the coelom. Three regions differentiate, a dorsal *epimere*, a central *mesomere* or *nephrotome*, and a ventral *hypomere* or *lateral plate*, the latter splitting into splanchnic and somatic layers as above. The *EPIMERE* segments from before backwards, to give a series of *somites* separated from the rest of the mesoderm. Within each somite, mid-wall cells migrate as *sclerotome* to the notochord, to form the vertebral column; outer-wall cells migrate as *dermatome* to form the dermis; and the rest remains in place as *myotome*, giving the voluntary segmental muscles. The *NEPHROTOME* gives rise to the kidneys and gonads and their ducts. The *HYPOMERE* has as its cavity the adult coelom, splanchnic material giving smooth muscle and connective tissue for the digestive tract and its outgrowths; also the heart develops from this area, as do the visceral serosa and the mesenteries.

MESENCHYME is a kind of primitive connective tissue, derived largely but not entirely from mesoderm. It gives rise to all the connective tissues of the body, to all smooth muscle, to blood cells and vessels, to lymph vessels and glands, and to the voluntary muscles of the limbs.

Adaptations of the Embryo.

Not all embryonic organs are precursors of adult organs, but some have arisen to facilitate feeding, or oxygen uptake, or protection of the embryo alone. In meroblastic eggs of birds, the yolk comes to be enclosed in a YOLK SAC. In chick, for example, the area opaca becomes vascularised as the area vasculosa, the first blood cells developing in its blood islands. The network is connected to the embryo, and the heart beats as early as the second day of incubation (total incubation is twenty-one days). Vascular downgrowths go deep into the yolk from the yolk sac wall for efficient absorption, and eventually the embryonic body is separated from the yolk sac by body folds growing under it, leaving it joined only by a cord, the sac being an adaptation for the removal of a lot of food in a short time. Besides the yolk-sac, three other extra-embryonic membranes are developed. These are the amnion, chorion, and allantois. In chick, the amniotic folds grow up over the embryo and then fuse to enclose it in an amniotic cavity, which becomes fluid-filled. The AMNION is the membrane lining this cavity, while the CHORION is the outer surface of the fold away from the embryo, and continuous with the epithelium of the yolk sac. In this way the embryo becomes immersed in a liquid medium, even though the egg as a whole is terrestrial. Besides avoiding dessication, the liquid is a shock-absorber, and also isolates the embryo from the shell, protecting it from abrasion and adhesion.

The ALLANTOIS is of very different nature, being functionally a urinary bladder. A cleidoic egg, being a closed system, cannot excrete waste nitrogen outside itself, and so it must be stored as insoluble uric acid in the allantois inside the egg. The allantois grows as a ventral evagination of the endodermal hindgut, and spreads rapidly through the extra-embryonic coelom, until by the eleventh day it covers the entire circumference of the egg below the chorion. At this stage its surface area is so large that it can assume a new function – that of gaseous exchange. A network of blood vessels is established, and new connections made in the embryo, the allantoic circulation continuing until just before hatching, when the allantois is nipped off and left behind. In this way the lessening of oxygen supplies occasioned by the embryo's sinking away from the surface of the egg into the amniotic cavity is more than compensated.

MAMMALIAN DEVELOPMENT is greatly modified compared with other vertebrates. Originally, they must have laid meroblastic eggs, as platypus still does; while in the marsupials the eggs are retained in the uterus for a short while, although the young are born in very immature condition. Here the reduced yolk is not used, but is ejected from the egg at the beginning of cleavage. In true placental mammals the egg has no yolk from the start, and cleavage is complete but irregular. The irregularity is due to a very early specialisation of the cells, cleavage giving a central group of cells suspended within a ball of superficial cells, the two kinds having very different prospective significances. This *blastocyst* corresponds anatomically to the blastula and in it there appears a layer of flat cells, the *hypoblast*, corresponding to the lower layer of cells of the chick blastoderm, and representing most of the presumptive endoderm. The rest of the inner cell mass is then equivalent to the ectoderm. The outer layer of cells, the *trophoblast*, enclosing the mass of cells makes no further contribution to the embryo as such, and corresponds in position to the chick chorion, but arises in a very different way. This is so because its function is to establish connection between mother and embryo, which must be done very early in development because the egg has no yolk. Gastrulation follows, with a primitive streak developing and fading as in chick, mesoderm being formed in the same way, and subsequent organogany being basically the same as in other vertebrates. The trophoblast is an adaptation to food-getting even prompter in appearance than the yolk-sac of birds. The extra-embryonic parts are of very great significance in mammals. When the blastocyst reaches the uterus it implants the trophoblast cells eroding the epithelial lining at the place of contact. The development of the *allantois* is highly significant, since with viviparity its function as a urinary bladder is superfluous, as urea is formed and quickly excreted via the mother. On the other hand, its originally secondary function of gaseous exchange is now vital, and is even extended to include food supply in the absence of yolk from the egg. The endodermal part of the allantois is reduced, but the blood-forming mesodermal component develops progressively. The true mammalian placenta is the CHORIO-ALLANTOIC PLACENTA, composed of maternal and foetal tissues jointly, serving the function of an exchange between maternal and foetal blood components, without allowing the bloods actually to mix. Obviously the efficiency of the interchange is increased both by increasing the area of contact by villi, and by reducing the thickness of the diffusion-path between the bloods by removing some of the original six layers. Both methods have evolved, the result being that mammals are born in an advanced state of readiness to face the rigours of the external environment.

An Atlas of Embryology

A. **Amphioxus:** cleavage, 1-cell stage, E. *mag. 450*×

B. **Amphioxus:** cleavage, 2-cell stage, E. *mag. 450*×

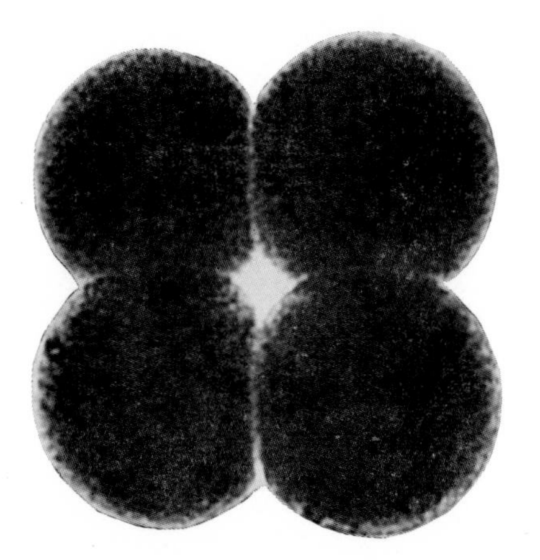

C. **Amphioxus:** cleavage, 4-cell stage, E. *mag. 450*×

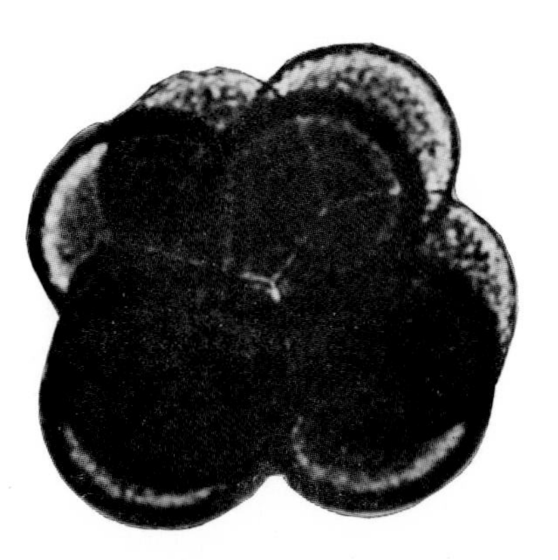

D. **Amphioxus:** cleavage, 8-cell stage, E. *mag. 450*×

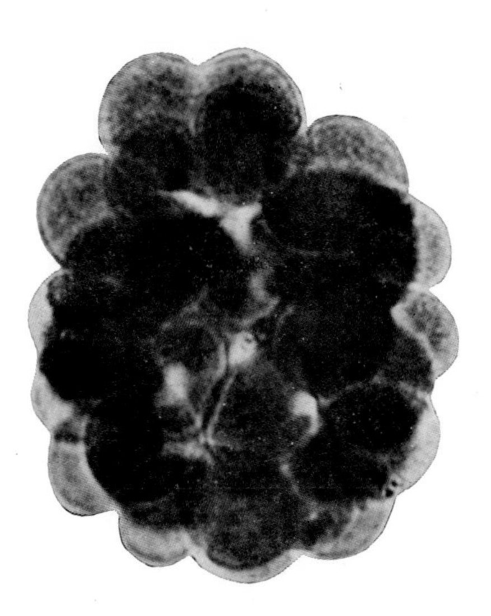

E. **Amphioxus:** cleavage, morula, E. *mag. 450×*

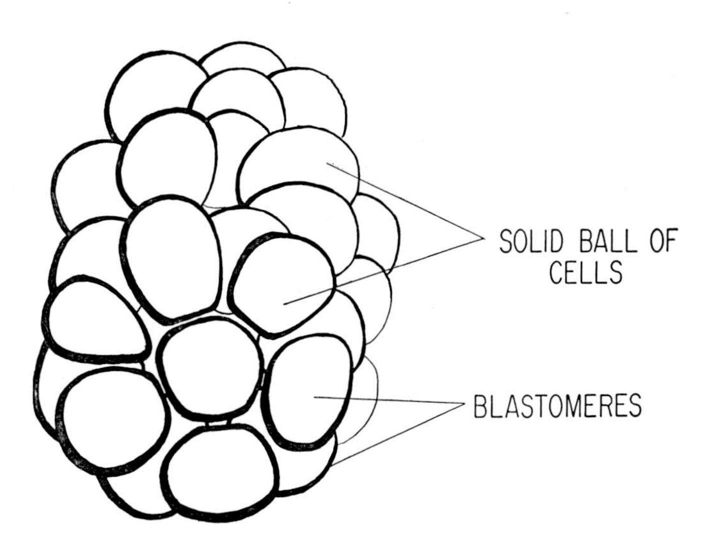
SOLID BALL OF CELLS
BLASTOMERES

E. Diagram of a Morula based on Specimen E.

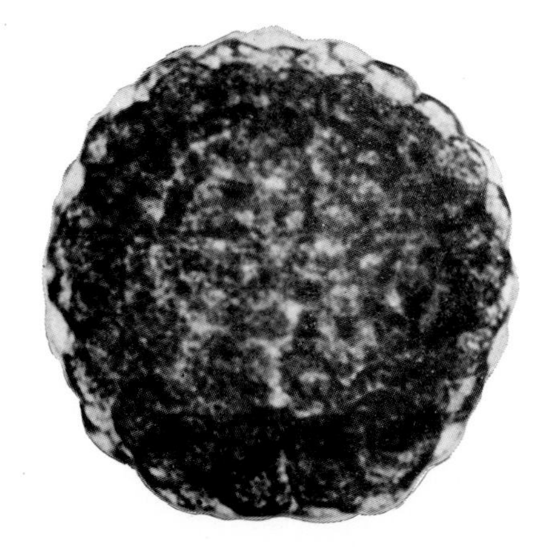

F. **Amphioxus:** cleavage, blastula, E. *mag. 450×*

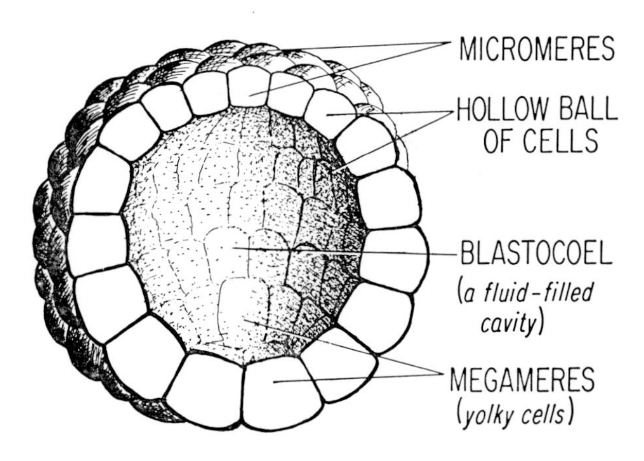
MICROMERES
HOLLOW BALL OF CELLS
BLASTOCOEL
(*a fluid-filled cavity*)
MEGAMERES
(*yolky cells*)

F. Diagram of Blastula, based on Specimen F.

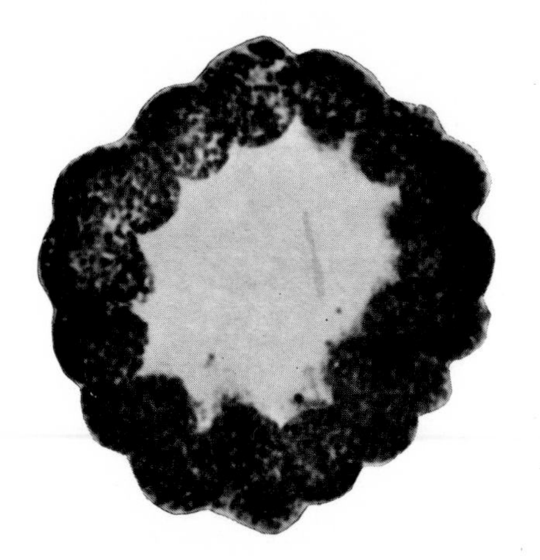

G. **Amphioxus:** cleavage, blastula, V.S. *mag. 450×*

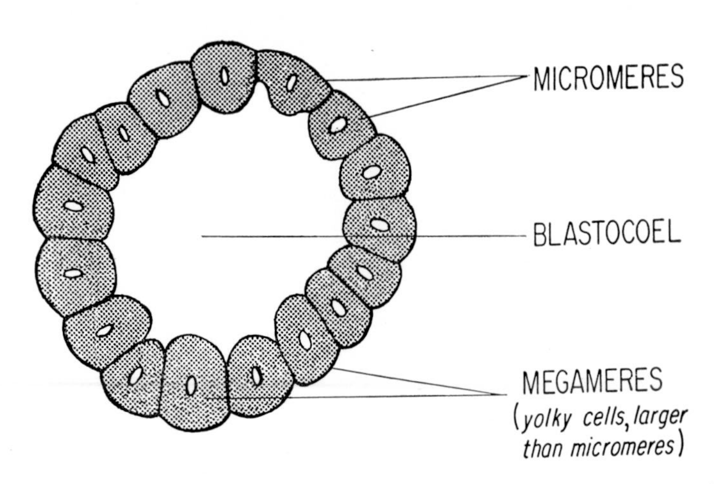
MICROMERES
BLASTOCOEL
MEGAMERES
(yolky cells, larger than micromeres)

G. Drawing of a V.S. of a blastula based on Specimen G.

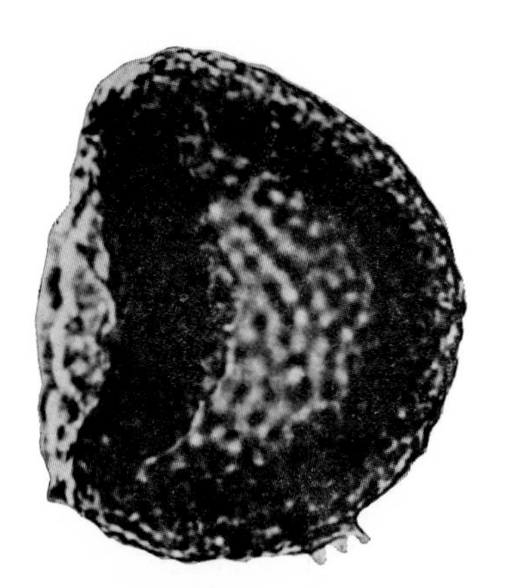

H. **Amphioxus:** gastrula, E. *mag. 450×*

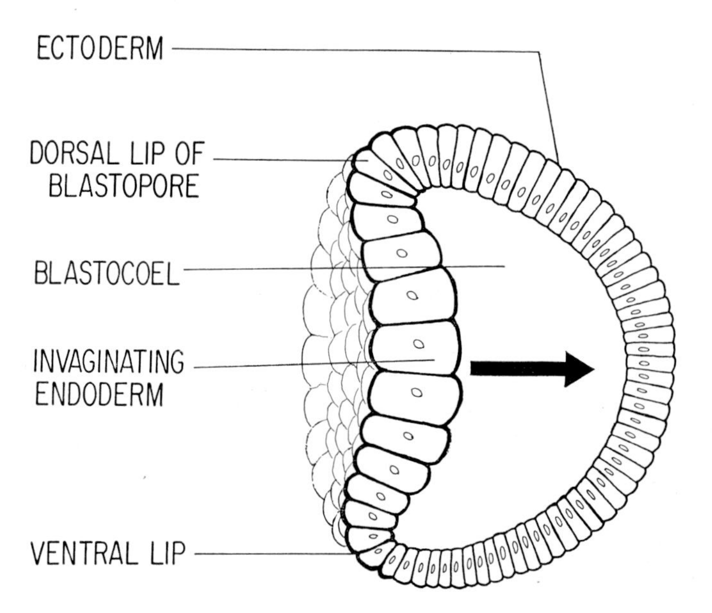
ECTODERM
DORSAL LIP OF BLASTOPORE
BLASTOCOEL
INVAGINATING ENDODERM
VENTRAL LIP

H. Diagram of Gastrula, based on Specimen H.

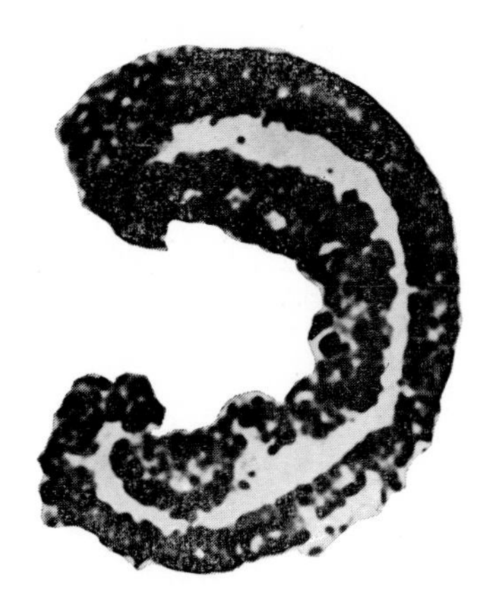

J. **Amphioxus:** gastrula, V.S. *mag. 450×*

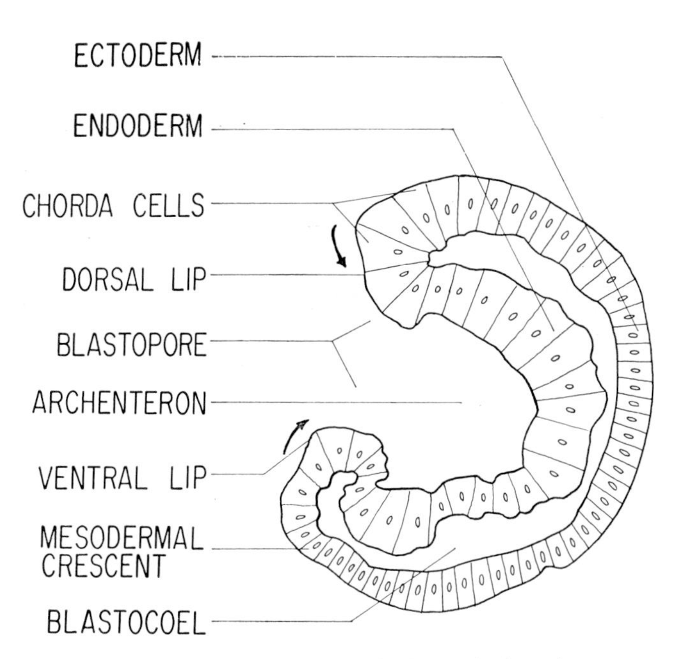
ECTODERM
ENDODERM
CHORDA CELLS
DORSAL LIP
BLASTOPORE
ARCHENTERON
VENTRAL LIP
MESODERMAL CRESCENT
BLASTOCOEL

J. Diagram of Section through Gastrula, based on Specimen J.

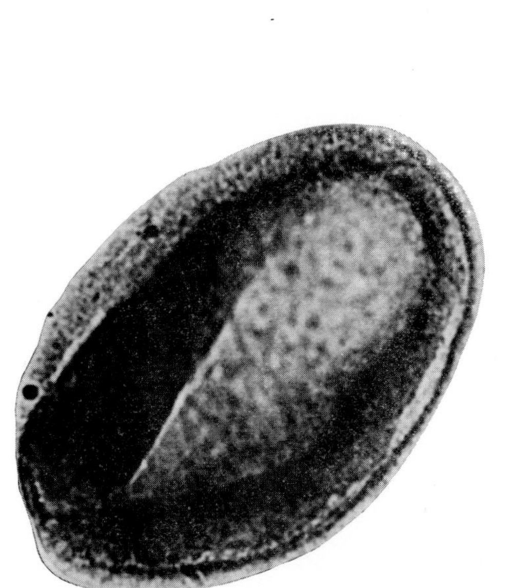

K. **Amphioxus:** early embryo, E. *mag. 450×*

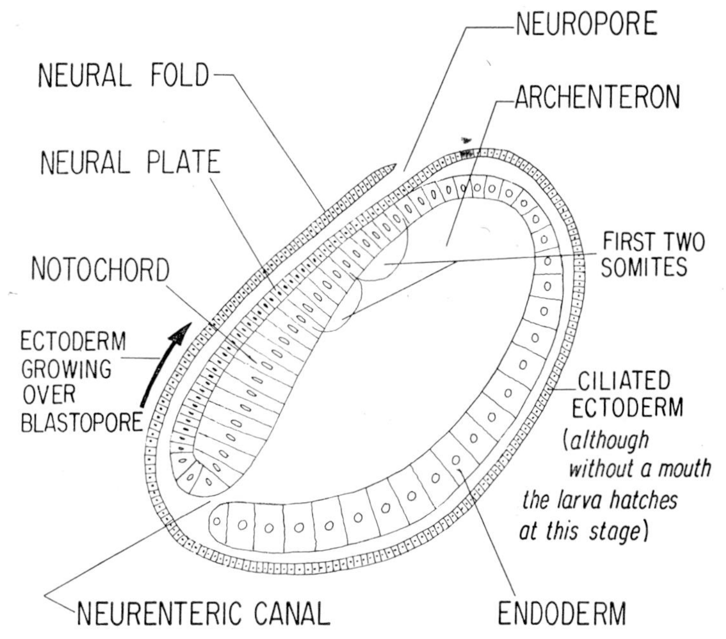
NEUROPORE
NEURAL FOLD
ARCHENTERON
NEURAL PLATE
FIRST TWO SOMITES
NOTOCHORD
ECTODERM GROWING OVER BLASTOPORE
CILIATED ECTODERM
(although without a mouth the larva hatches at this stage)
NEURENTERIC CANAL
ENDODERM

K. Diagram of early embryo based on Specimen K.

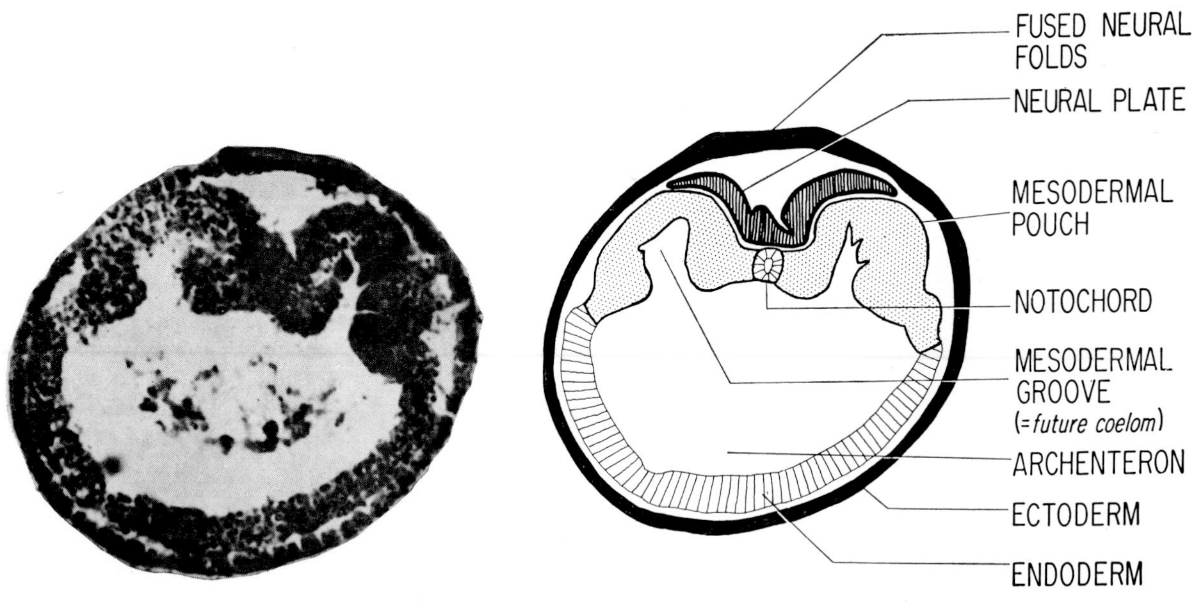
FUSED NEURAL FOLDS
NEURAL PLATE
MESODERMAL POUCH
NOTOCHORD
MESODERMAL GROOVE (=*future coelom*)
ARCHENTERON
ECTODERM
ENDODERM

L. **Amphioxus:** 8-somite stage, T.S. *mag. 450×*

L. Diagram of a T.S. of a 8-somite stage based on Specimen L.

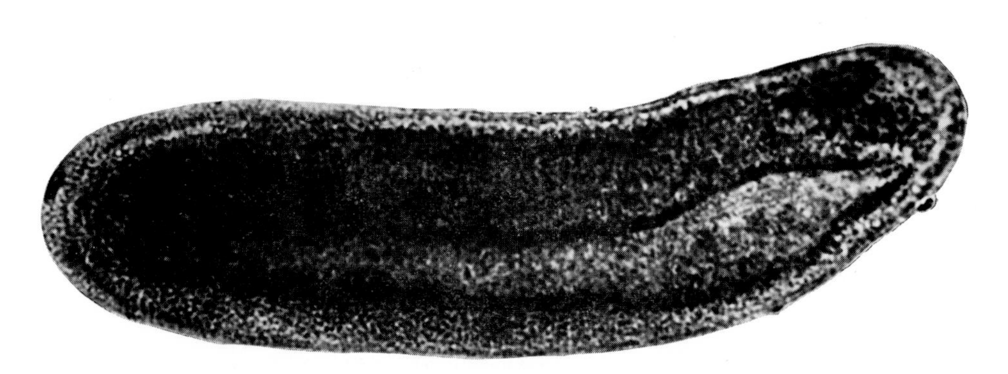

M. **Amphioxus:** 12-somite stage, E. *mag. 420×*

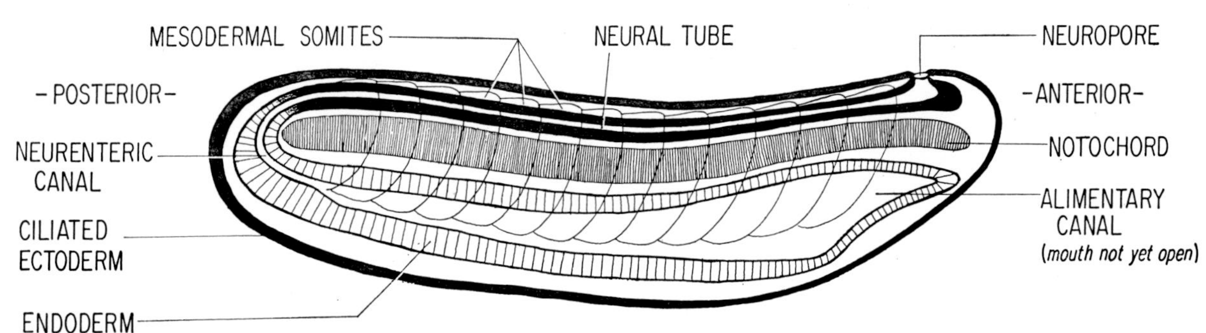
MESODERMAL SOMITES
NEURAL TUBE
NEUROPORE
- POSTERIOR -
- ANTERIOR -
NEURENTERIC CANAL
NOTOCHORD
CILIATED ECTODERM
ALIMENTARY CANAL
(*mouth not yet open*)
ENDODERM

M. Diagram of 12-somite stage based on Specimen M.

N. **Amphioxus:** 17-somite stage, E. *mag. 420×*

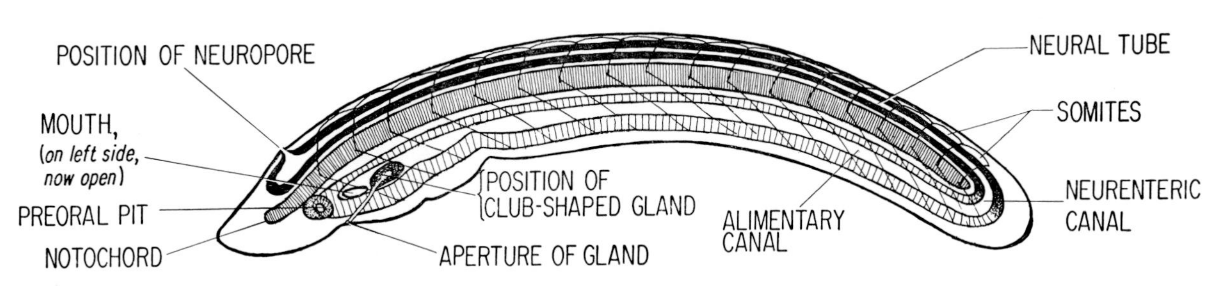
POSITION OF NEUROPORE
MOUTH,
(on left side, now open)
PREORAL PIT
NOTOCHORD
POSITION OF CLUB-SHAPED GLAND
APERTURE OF GLAND
ALIMENTARY CANAL
NEURAL TUBE
SOMITES
NEURENTERIC CANAL

N. Diagram of a 17 somite stage based on Specimen N.

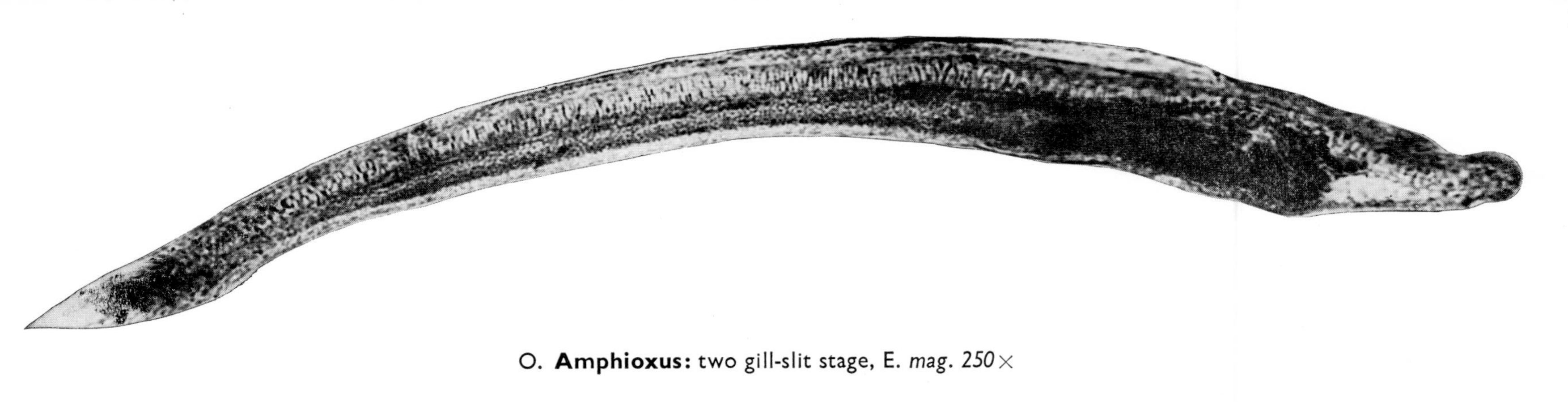

O. **Amphioxus:** two gill-slit stage, E. *mag.* 250×

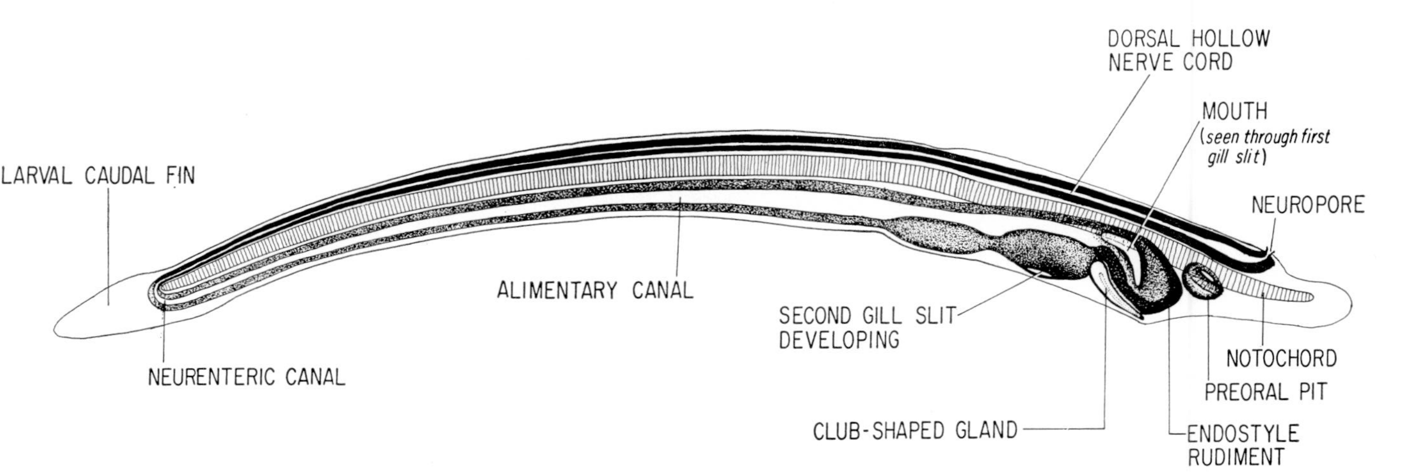
DORSAL HOLLOW NERVE CORD
MOUTH
(*seen through first gill slit*)
NEUROPORE
LARVAL CAUDAL FIN
ALIMENTARY CANAL
SECOND GILL SLIT DEVELOPING
NOTOCHORD
PREORAL PIT
NEURENTERIC CANAL
CLUB-SHAPED GLAND
ENDOSTYLE RUDIMENT

O. Diagram of two gill slit stage based on Specimen O.

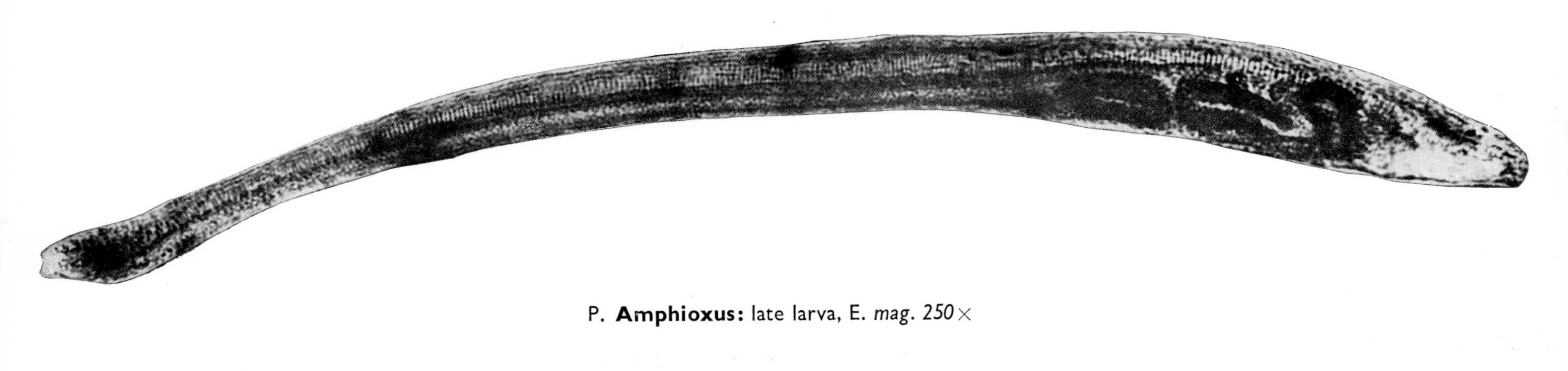

P. **Amphioxus:** late larva, E. *mag.* 250×

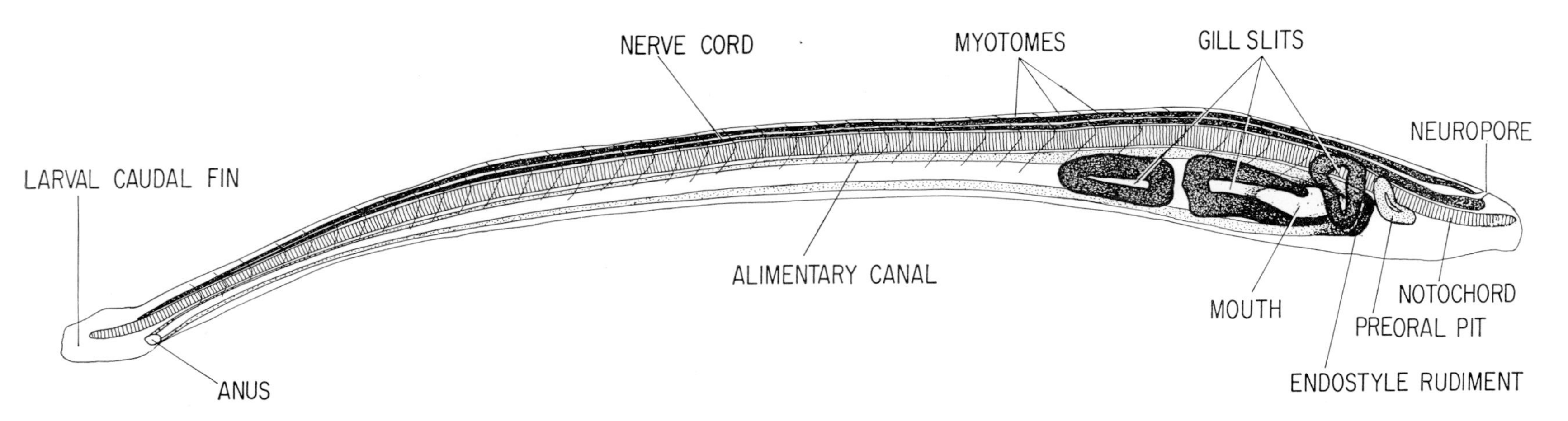
NERVE CORD
MYOTOMES
GILL SLITS
NEUROPORE
LARVAL CAUDAL FIN
ALIMENTARY CANAL
MOUTH
NOTOCHORD
PREORAL PIT
ENDOSTYLE RUDIMENT
ANUS

P. Diagram of 3 gill slit stage based on Specimen P.

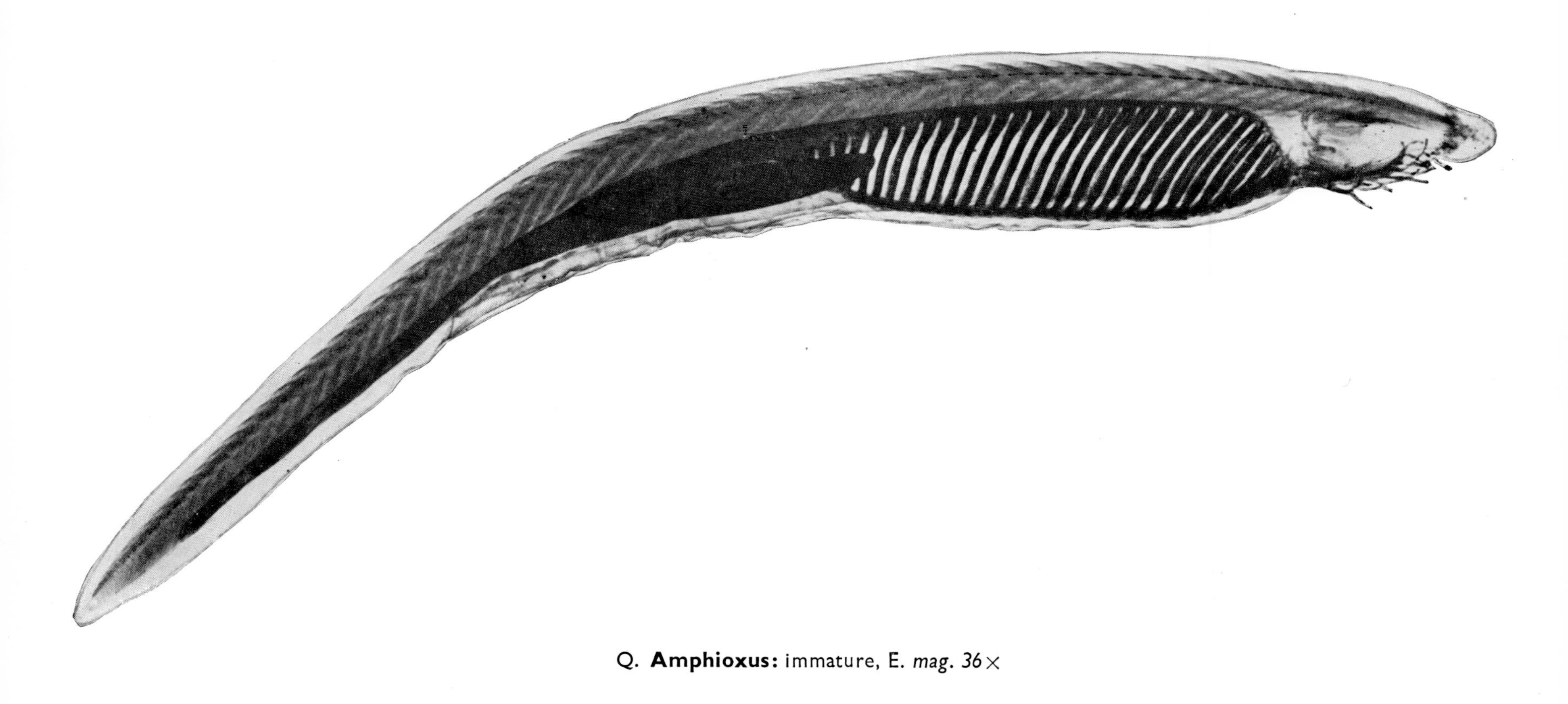

Q. **Amphioxus:** immature, E. *mag. 36*×

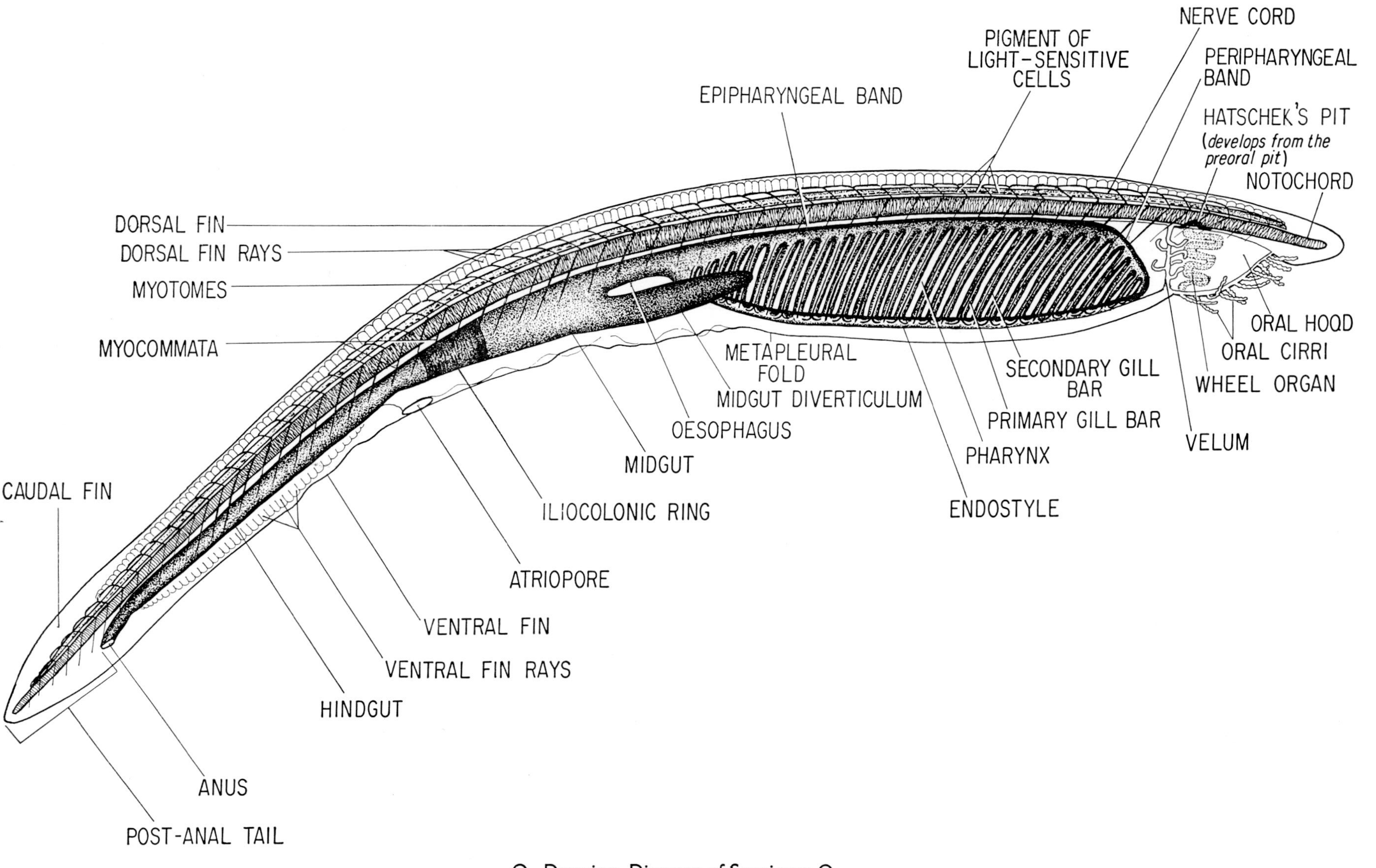
NERVE CORD
PIGMENT OF LIGHT-SENSITIVE CELLS
PERIPHARYNGEAL BAND
EPIPHARYNGEAL BAND
HATSCHEK'S PIT (develops from the preoral pit)
NOTOCHORD
DORSAL FIN
DORSAL FIN RAYS
MYOTOMES
MYOCOMMATA
ORAL HOOD
ORAL CIRRI
WHEEL ORGAN
METAPLEURAL FOLD
SECONDARY GILL BAR
MIDGUT DIVERTICULUM
PRIMARY GILL BAR
VELUM
OESOPHAGUS
PHARYNX
MIDGUT
CAUDAL FIN
ENDOSTYLE
ILIOCOLONIC RING
ATRIOPORE
VENTRAL FIN
VENTRAL FIN RAYS
HINDGUT
ANUS
POST-ANAL TAIL

Q. Drawing. Diagram of Specimen Q.

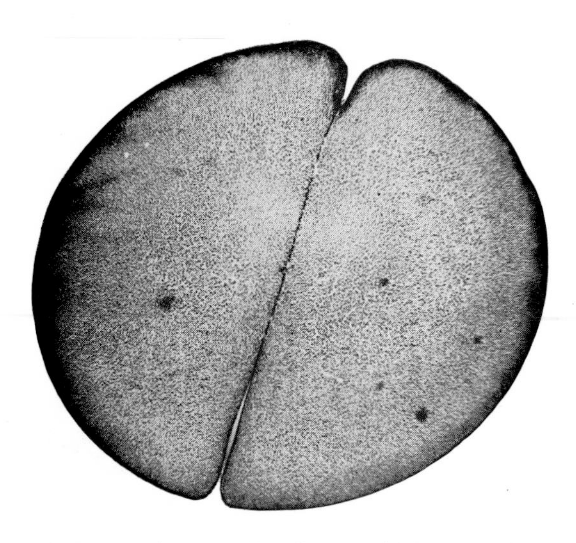

1. **Frog:** cleavage, 2-cell stage, V.S. *mag. 50*×

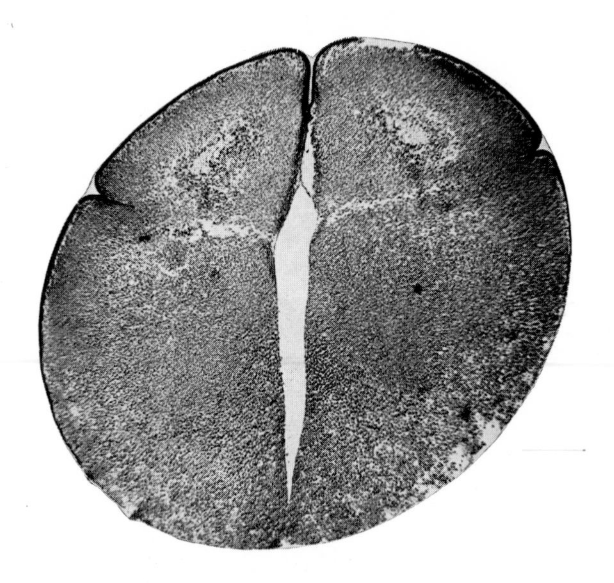

2. **Frog:** cleavage furrows, V.S. *mag. 50*×

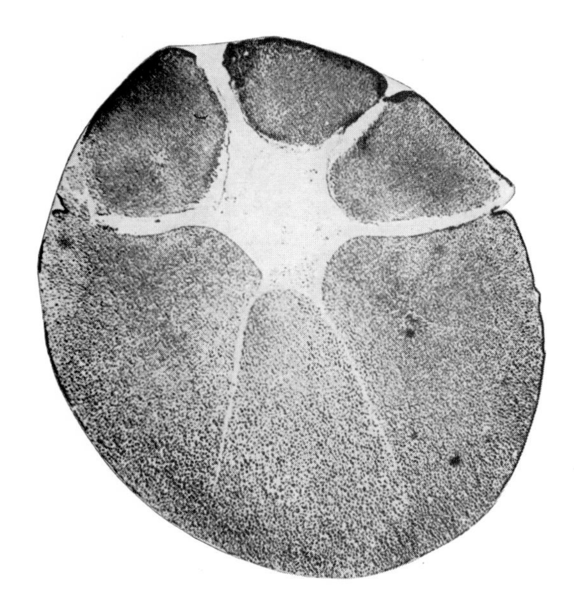

3. **Frog:** cleavage, 16-cell stage, V.S. *mag. 50*×

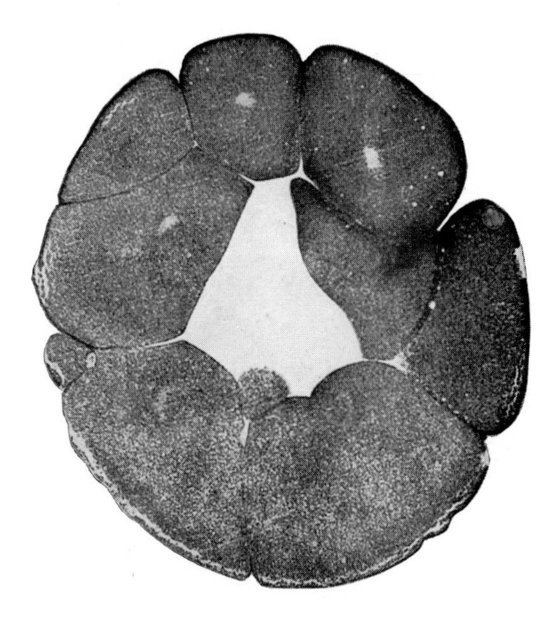

4. **Frog:** cleavage, 24-cell stage, V.S. *mag. 50*×

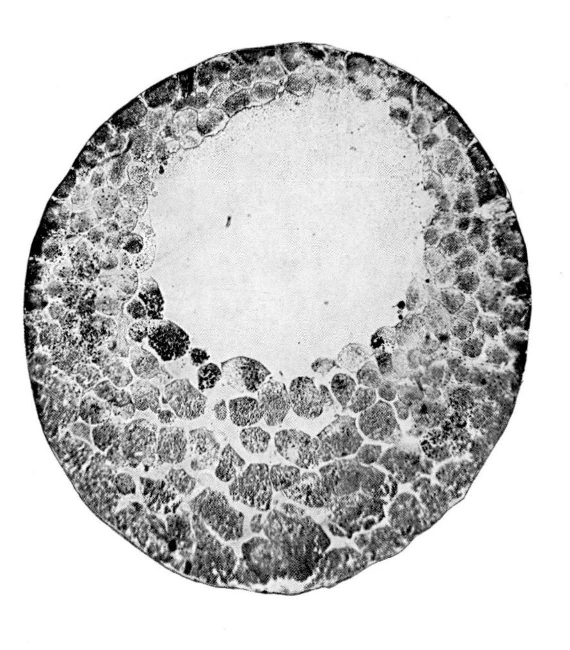

5. **Frog:** cleavage, blastula, V.S. *mag. 45* ×

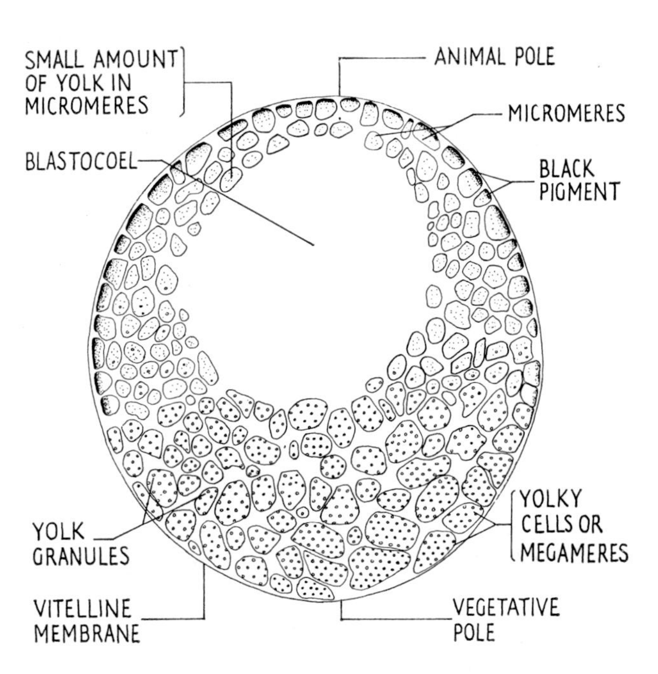
SMALL AMOUNT OF YOLK IN MICROMERES
ANIMAL POLE
MICROMERES
BLASTOCOEL
BLACK PIGMENT
YOLKY CELLS OR MEGAMERES
YOLK GRANULES
VITELLINE MEMBRANE
VEGETATIVE POLE

Drawing of specimen 5

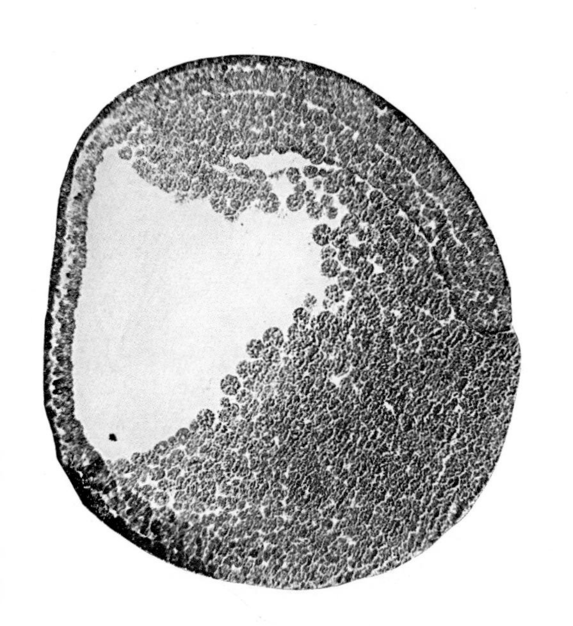

6. **Frog:** early gastrula (dorsal lip), V.S. *mag. 40* ×

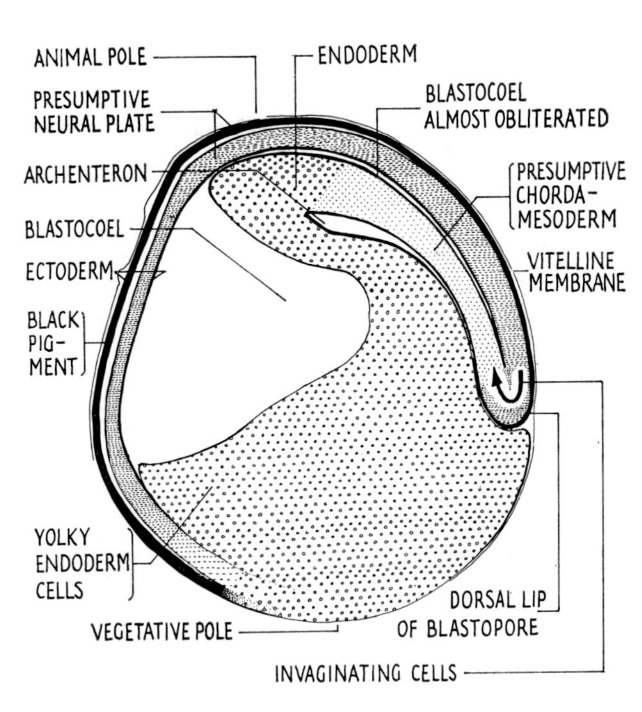
ANIMAL POLE
ENDODERM
PRESUMPTIVE NEURAL PLATE
BLASTOCOEL ALMOST OBLITERATED
ARCHENTERON
PRESUMPTIVE CHORDA-MESODERM
BLASTOCOEL
ECTODERM
VITELLINE MEMBRANE
BLACK PIG-MENT
YOLKY ENDODERM CELLS
DORSAL LIP OF BLASTOPORE
VEGETATIVE POLE
INVAGINATING CELLS

Drawing of specimen 6

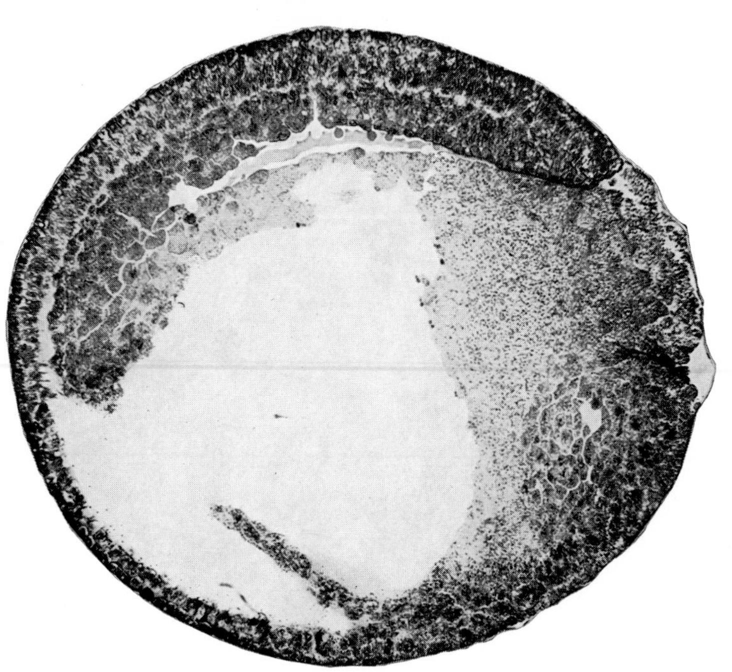

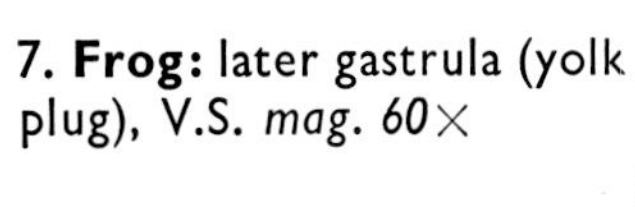

7. **Frog:** later gastrula (yolk plug), V.S. *mag. 60*×

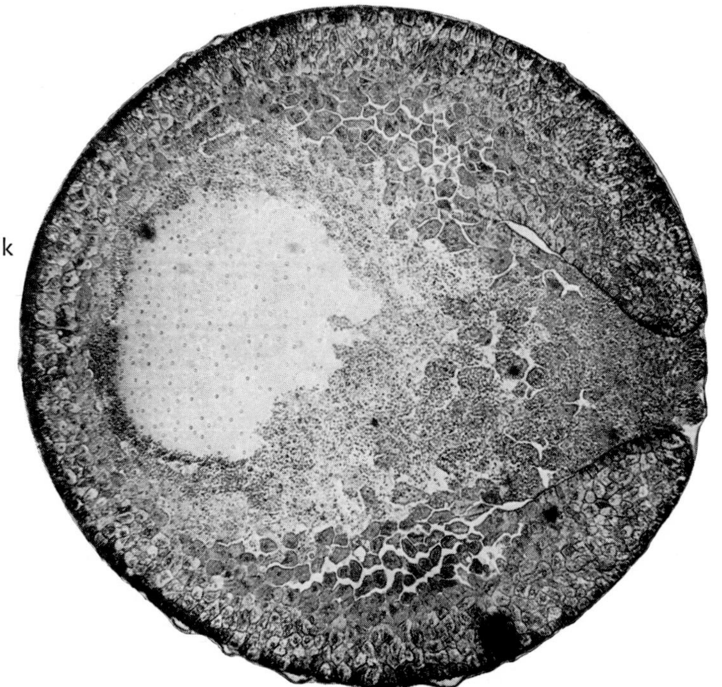

8. **Frog:** later gastrula (yolk plug), H.S. *mag. 60*×

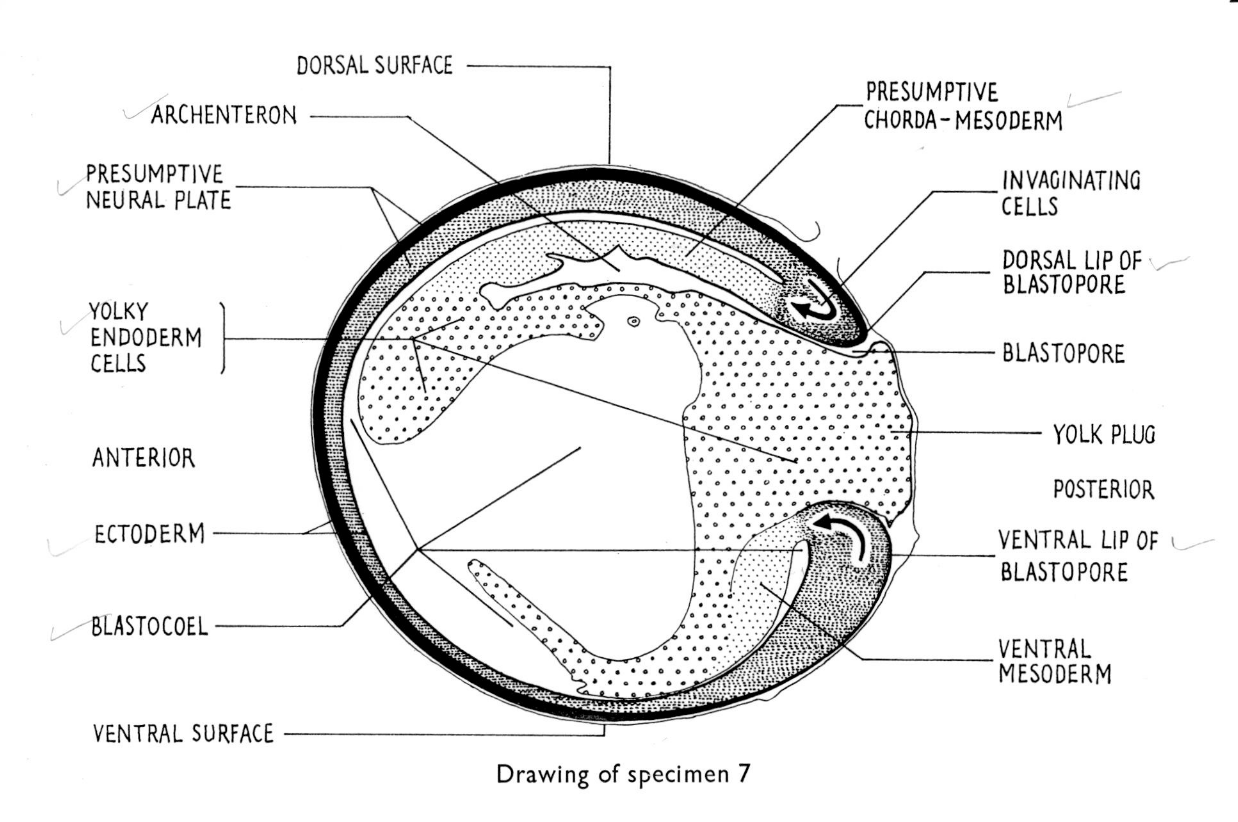
DORSAL SURFACE
ARCHENTERON
PRESUMPTIVE CHORDA-MESODERM
PRESUMPTIVE NEURAL PLATE
INVAGINATING CELLS
DORSAL LIP OF BLASTOPORE
YOLKY ENDODERM CELLS
BLASTOPORE
YOLK PLUG
ANTERIOR
POSTERIOR
ECTODERM
VENTRAL LIP OF BLASTOPORE
BLASTOCOEL
VENTRAL MESODERM
VENTRAL SURFACE

Drawing of specimen 7

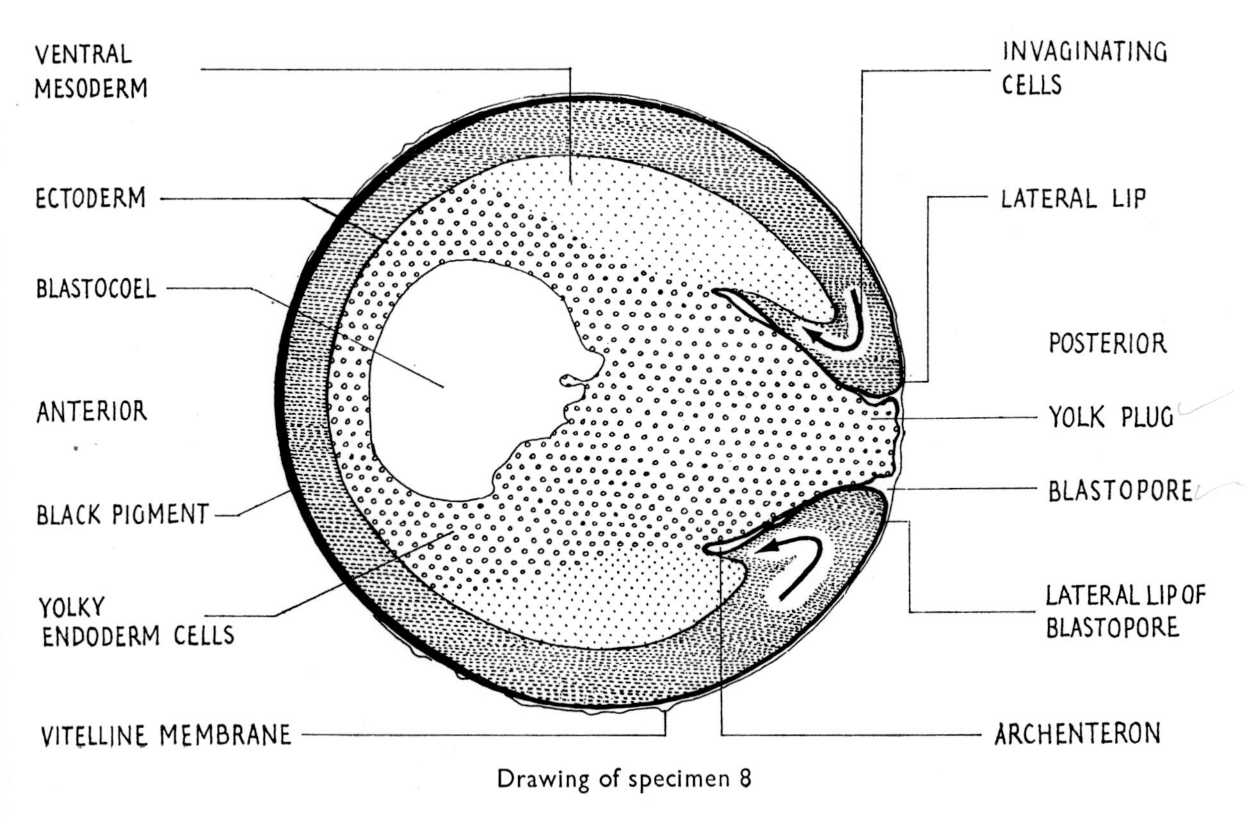
VENTRAL MESODERM
INVAGINATING CELLS
ECTODERM
LATERAL LIP
BLASTOCOEL
POSTERIOR
ANTERIOR
YOLK PLUG
BLASTOPORE
BLACK PIGMENT
YOLKY ENDODERM CELLS
LATERAL LIP OF BLASTOPORE
VITELLINE MEMBRANE
ARCHENTERON

Drawing of specimen 8

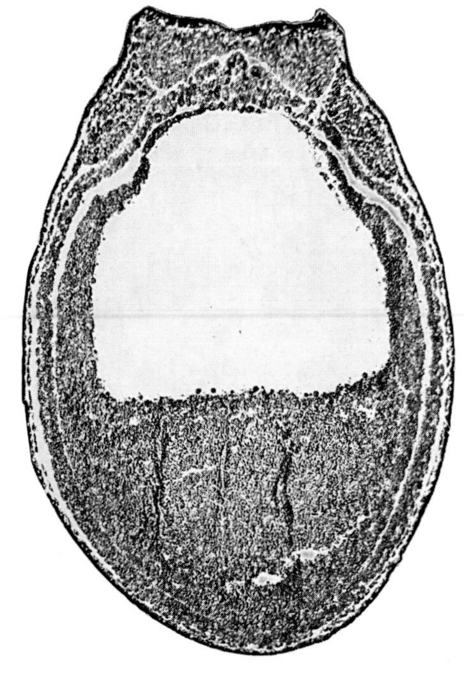

9. **Frog:** neural plate stage, T.S. *mag.* 35×

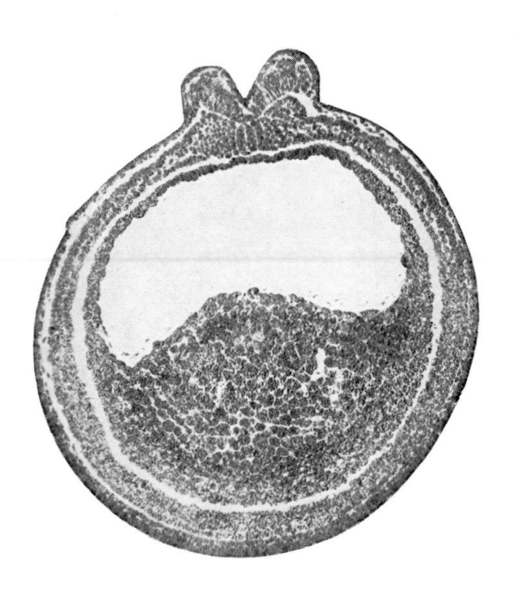

10. **Frog:** neural fold stage, T.S. *mag.* 35×

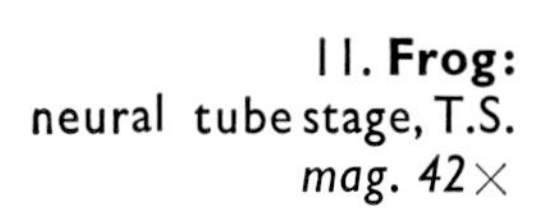

11. **Frog:** neural tube stage, T.S. *mag.* 42×

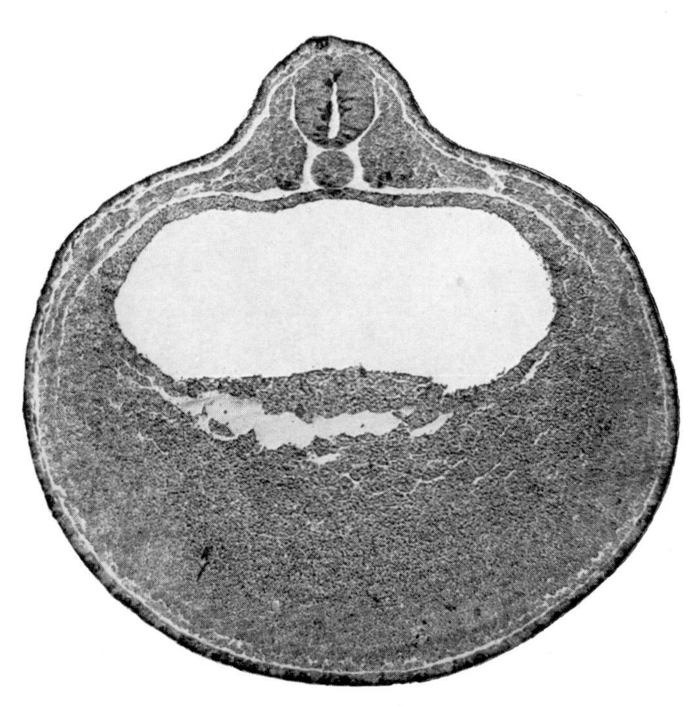

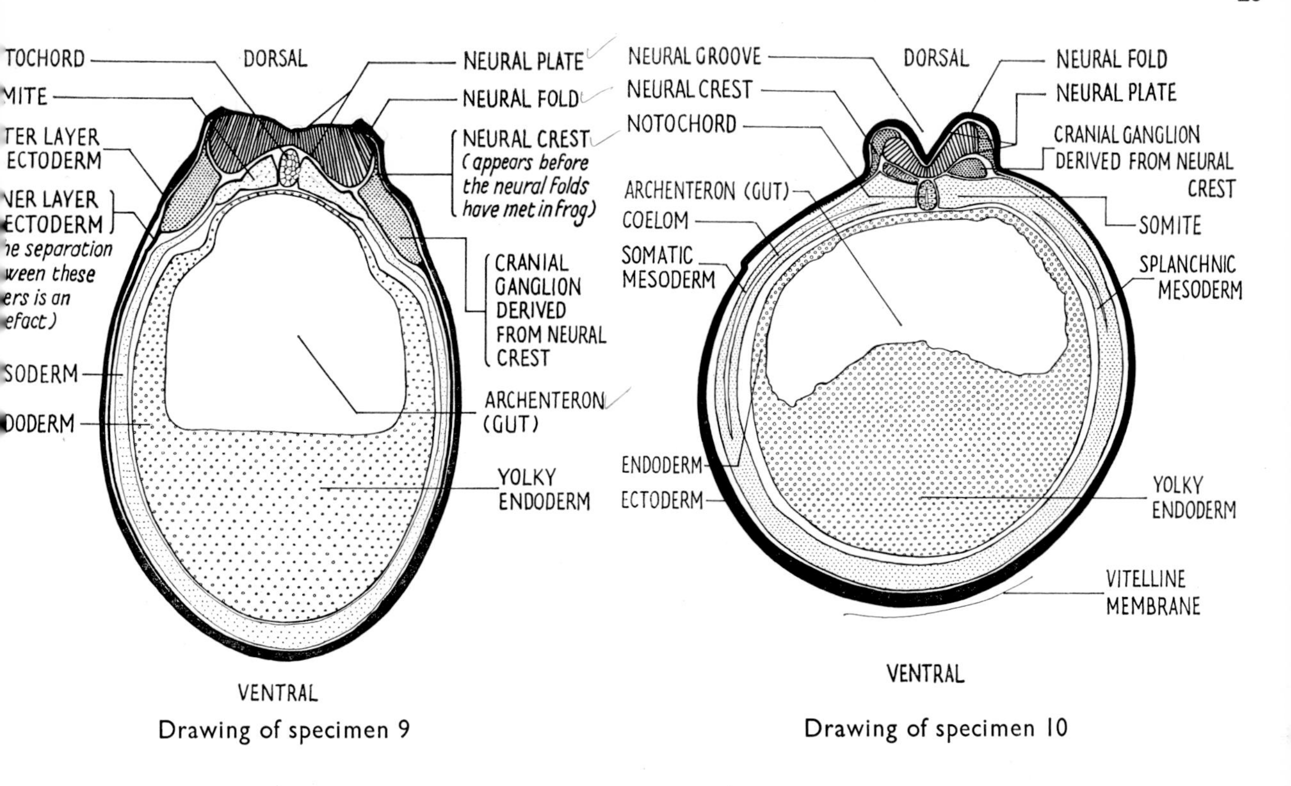
TOCHORD
DORSAL
NEURAL PLATE
MITE
NEURAL FOLD
TER LAYER
ECTODERM
NEURAL CREST
(appears before the neural folds have met in Frog)
NER LAYER
ECTODERM
he separation
ween these
ers is an
efact)
CRANIAL GANGLION DERIVED FROM NEURAL CREST
SODERM
ODERM
ARCHENTERON (GUT)
YOLKY ENDODERM
VENTRAL
NEURAL GROOVE
DORSAL
NEURAL FOLD
NEURAL CREST
NEURAL PLATE
NOTOCHORD
CRANIAL GANGLION DERIVED FROM NEURAL CREST
ARCHENTERON (GUT)
COELOM
SOMITE
SOMATIC MESODERM
SPLANCHNIC MESODERM
ENDODERM
ECTODERM
YOLKY ENDODERM
VITELLINE MEMBRANE
VENTRAL

Drawing of specimen 9

Drawing of specimen 10

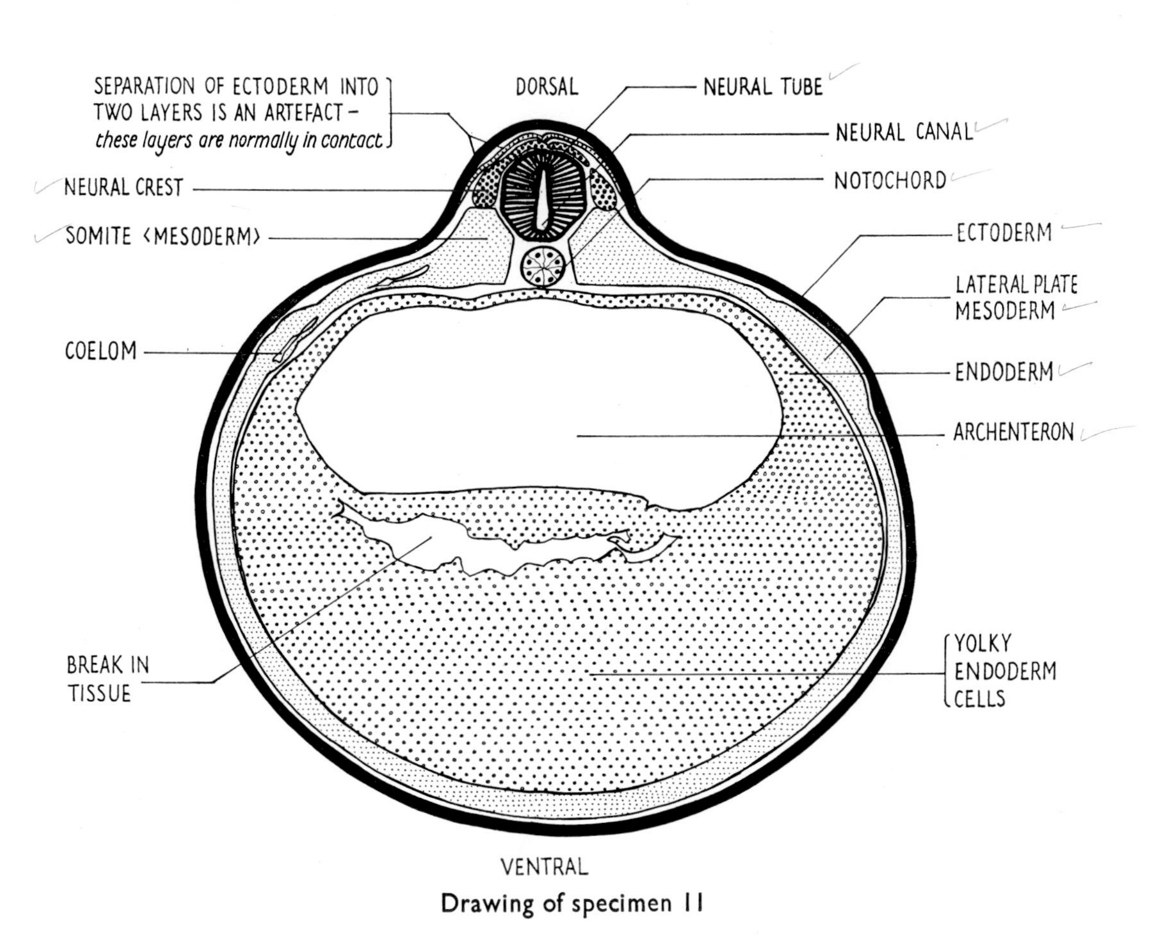
SEPARATION OF ECTODERM INTO TWO LAYERS IS AN ARTEFACT – these layers are normally in contact
DORSAL
NEURAL TUBE
NEURAL CANAL
NEURAL CREST
NOTOCHORD
SOMITE (MESODERM)
ECTODERM
LATERAL PLATE MESODERM
COELOM
ENDODERM
ARCHENTERON
BREAK IN TISSUE
YOLKY ENDODERM CELLS
VENTRAL

Drawing of specimen 11

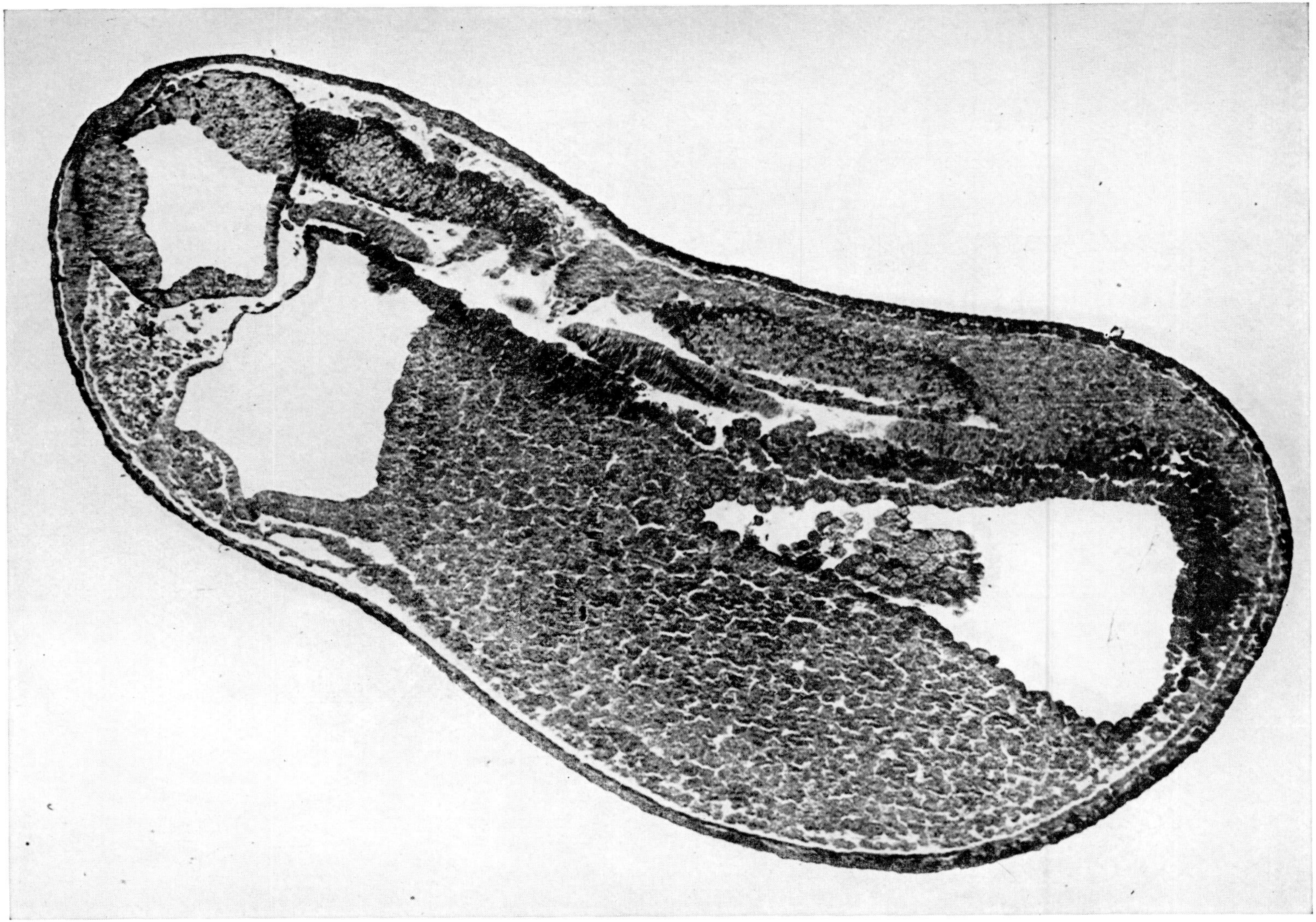

12. **Frog:** neurula, V.L.S. *mag. 60×*

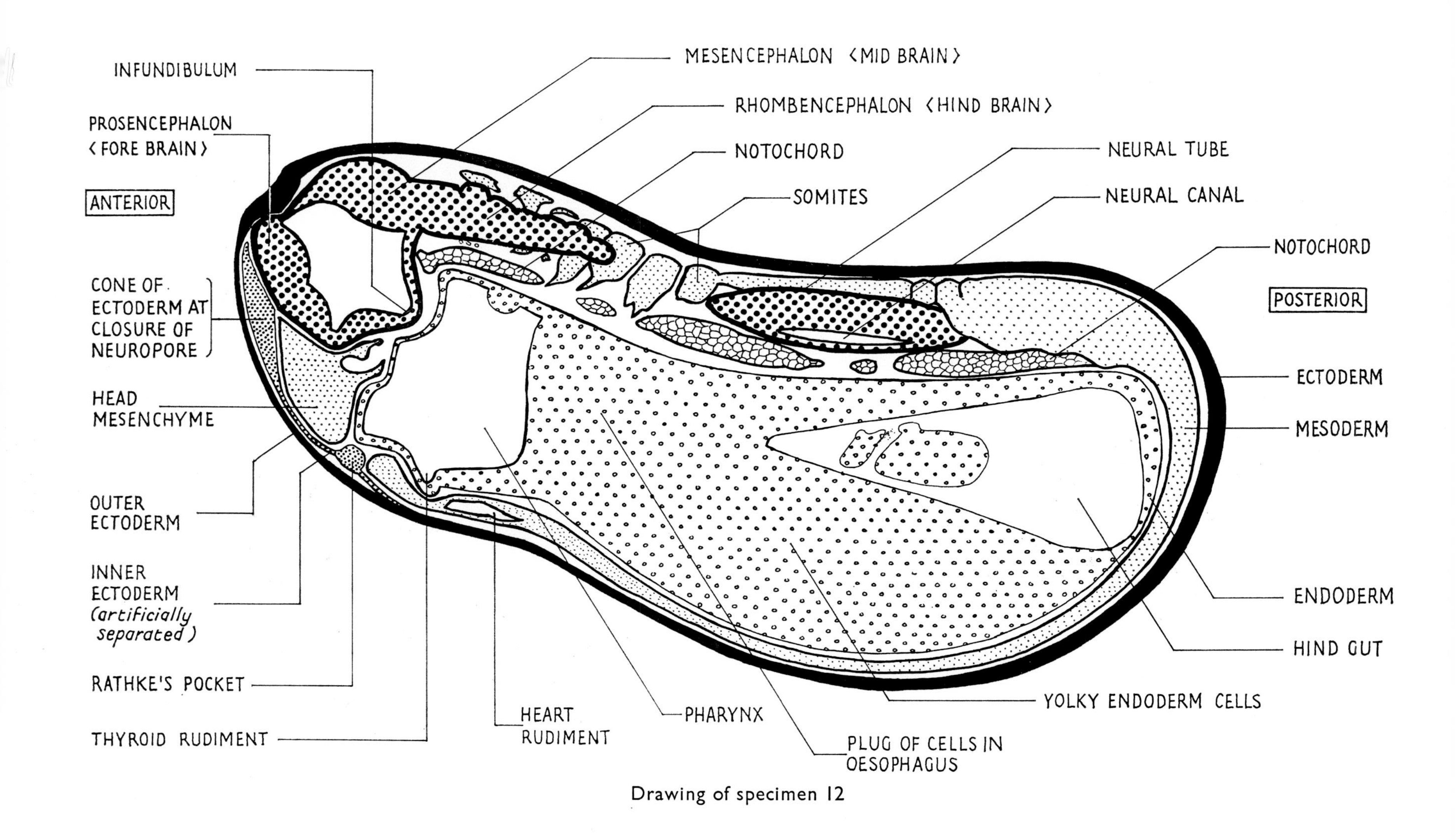

INFUNDIBULUM
PROSENCEPHALON
<FORE BRAIN>
ANTERIOR
CONE OF ECTODERM AT CLOSURE OF NEUROPORE
HEAD MESENCHYME
OUTER ECTODERM
INNER ECTODERM (artificially separated)
RATHKE'S POCKET
THYROID RUDIMENT
HEART RUDIMENT
PHARYNX
PLUG OF CELLS IN OESOPHAGUS
YOLKY ENDODERM CELLS
HIND GUT
ENDODERM
MESODERM
ECTODERM
POSTERIOR
NOTOCHORD
NEURAL CANAL
NEURAL TUBE
SOMITES
NOTOCHORD
RHOMBENCEPHALON <HIND BRAIN>
MESENCEPHALON <MID BRAIN>

Drawing of specimen 12

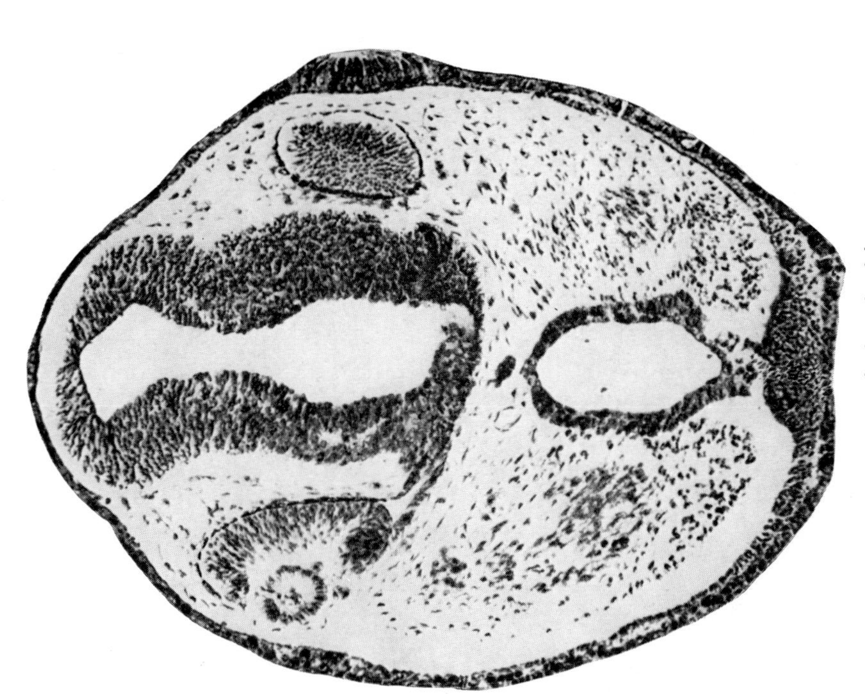

13. **Frog:** newly-hatched larva, optic region, T.S. *mag.* 80×

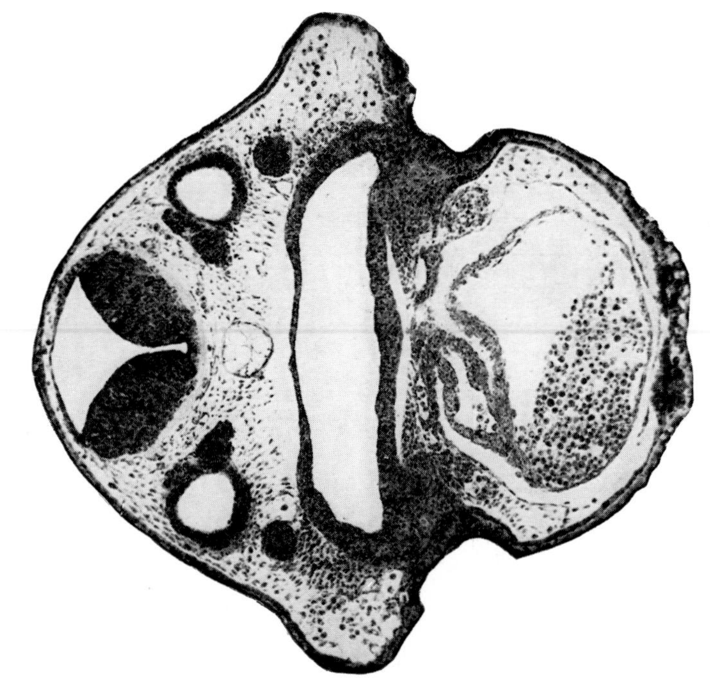

14. **Frog:** newly-hatched larva, auditory region, T.S. *mag.* 55×

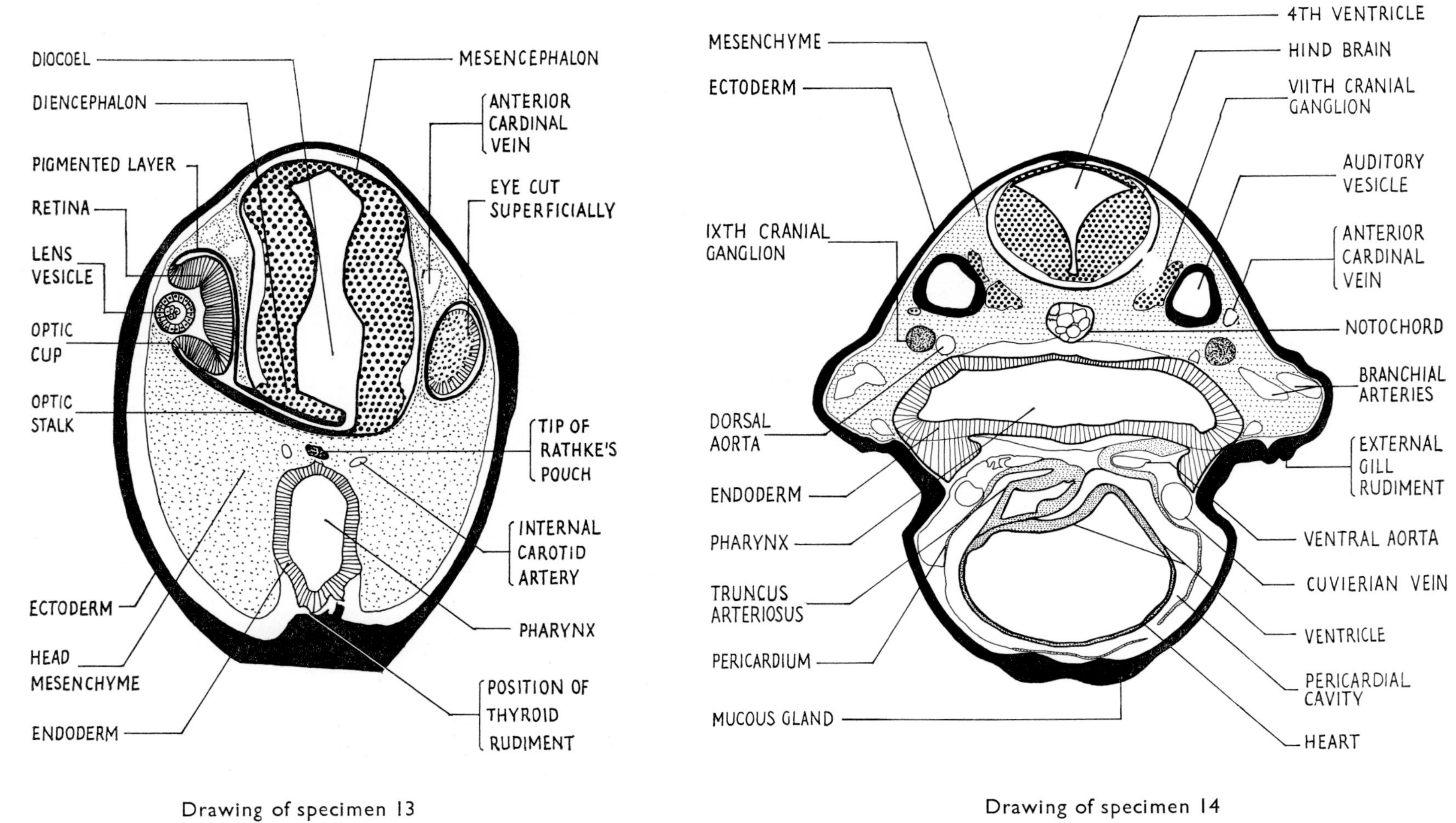
DIOCOEL
DIENCEPHALON
PIGMENTED LAYER
RETINA
LENS VESICLE
OPTIC CUP
OPTIC STALK
ECTODERM
HEAD MESENCHYME
ENDODERM
MESENCEPHALON
ANTERIOR CARDINAL VEIN
EYE CUT SUPERFICIALLY
TIP OF RATHKE'S POUCH
INTERNAL CAROTID ARTERY
PHARYNX
POSITION OF THYROID RUDIMENT
MESENCHYME
ECTODERM
IXTH CRANIAL GANGLION
DORSAL AORTA
ENDODERM
PHARYNX
TRUNCUS ARTERIOSUS
PERICARDIUM
MUCOUS GLAND
4TH VENTRICLE
HIND BRAIN
VIITH CRANIAL GANGLION
AUDITORY VESICLE
ANTERIOR CARDINAL VEIN
NOTOCHORD
BRANCHIAL ARTERIES
EXTERNAL GILL RUDIMENT
VENTRAL AORTA
CUVIERIAN VEIN
VENTRICLE
PERICARDIAL CAVITY
HEART

Drawing of specimen 13

Drawing of specimen 14

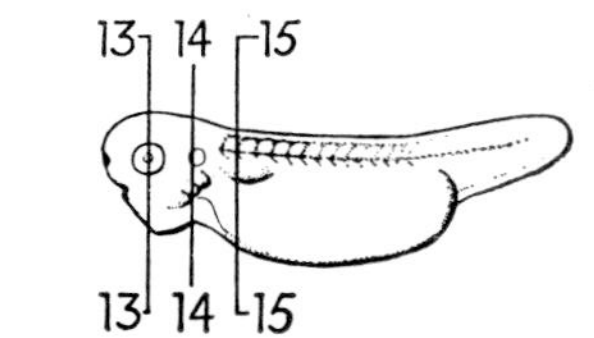
13
14
15
13
14
15

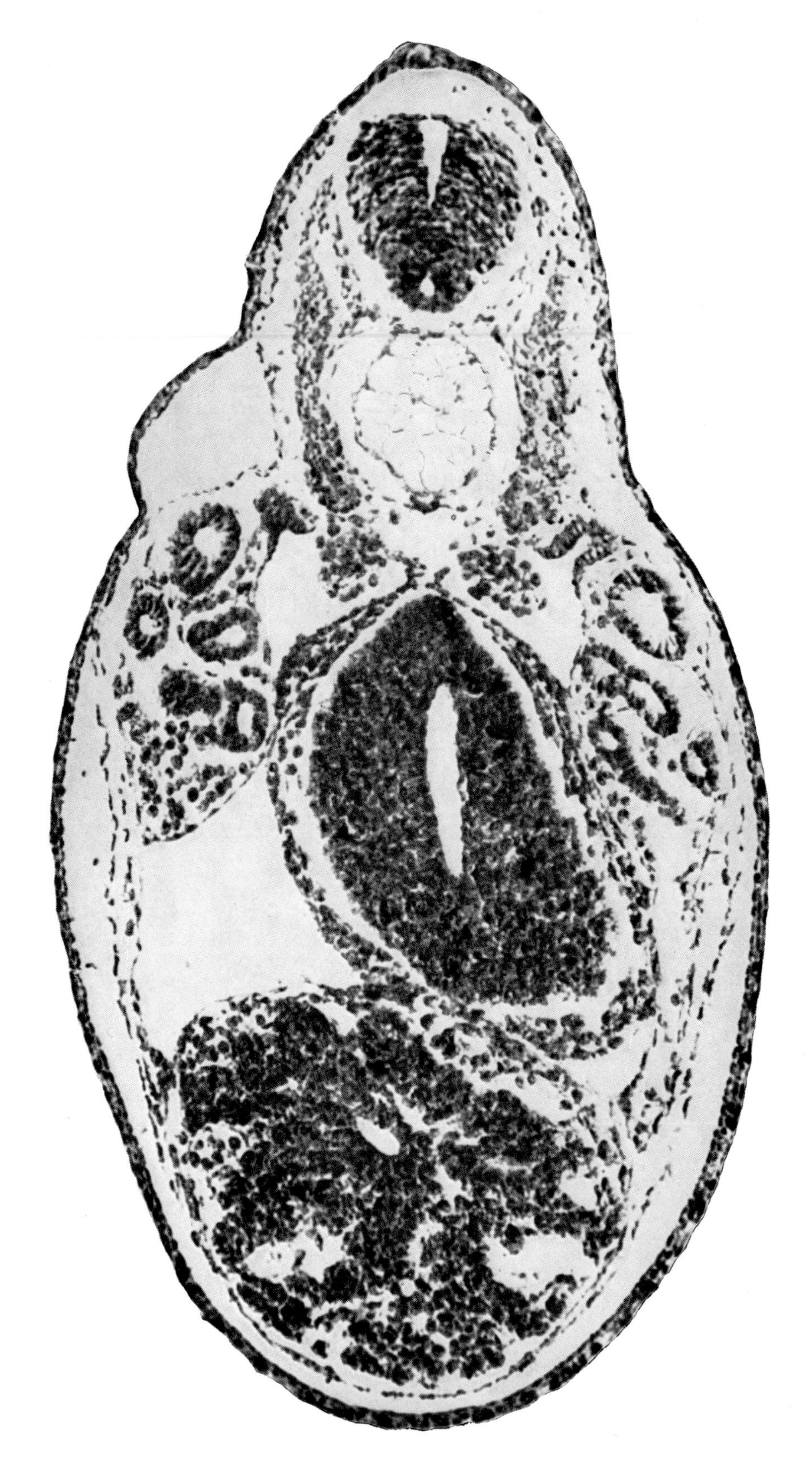

15. **Frog:** newly-hatched Larva, trunk region, T.S. *mag. 130×*

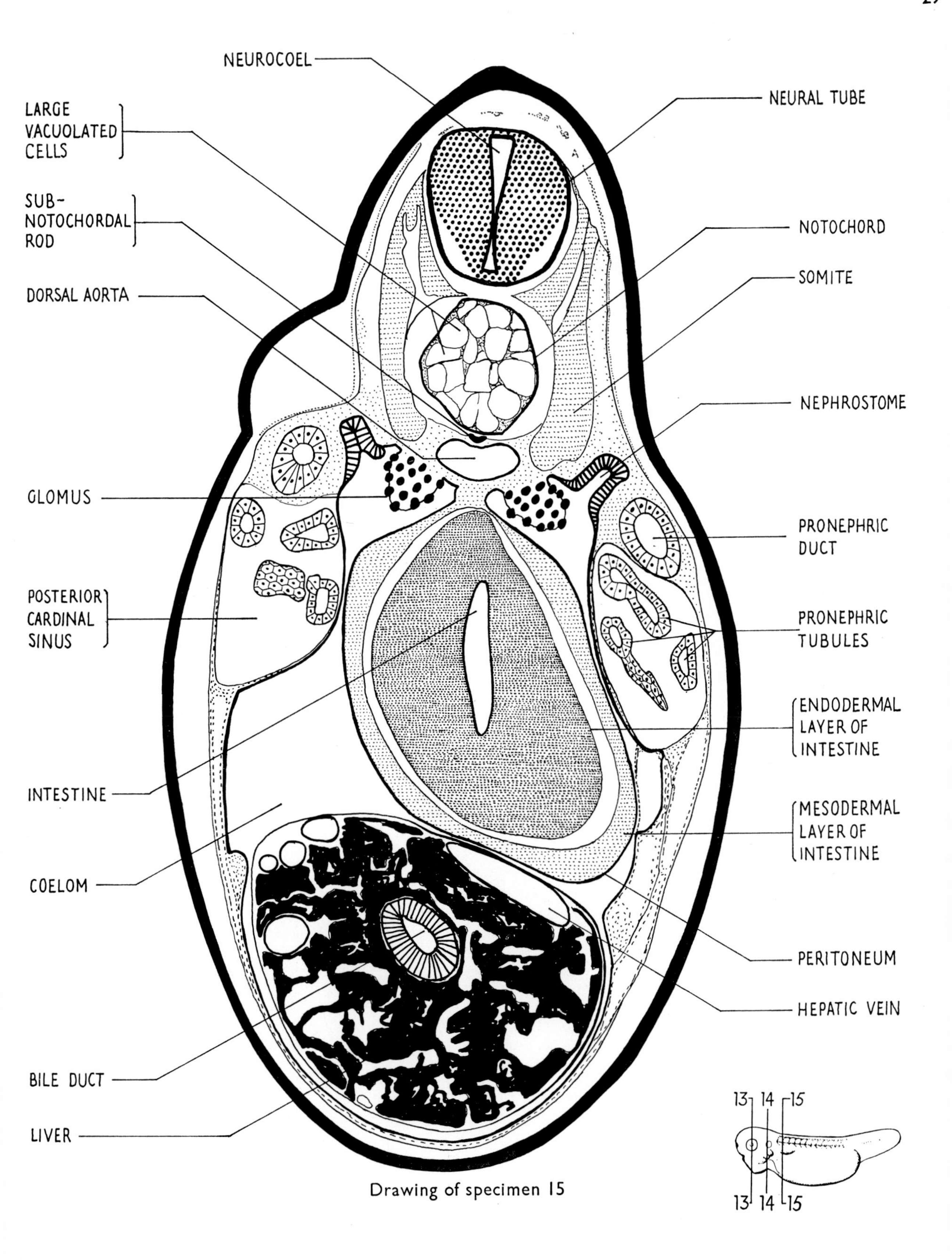
NEUROCOEL
LARGE VACUOLATED CELLS
SUB-NOTOCHORDAL ROD
DORSAL AORTA
GLOMUS
POSTERIOR CARDINAL SINUS
INTESTINE
COELOM
BILE DUCT
LIVER
NEURAL TUBE
NOTOCHORD
SOMITE
NEPHROSTOME
PRONEPHRIC DUCT
PRONEPHRIC TUBULES
ENDODERMAL LAYER OF INTESTINE
MESODERMAL LAYER OF INTESTINE
PERITONEUM
HEPATIC VEIN
13 14 15
13 14 15

Drawing of specimen 15

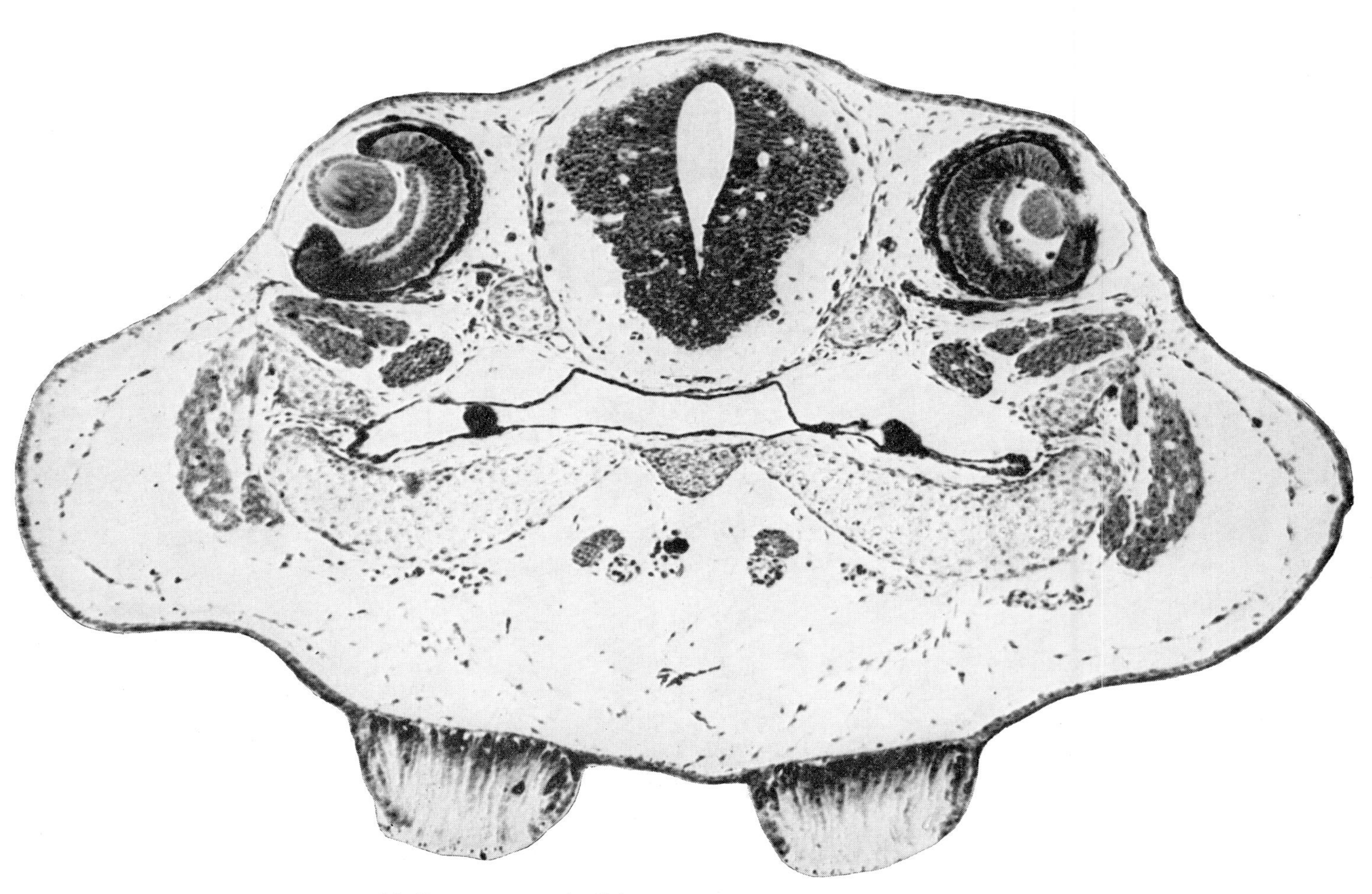

16. **Frog:** external gill larva, optic region, T.S. *mag. 100×*

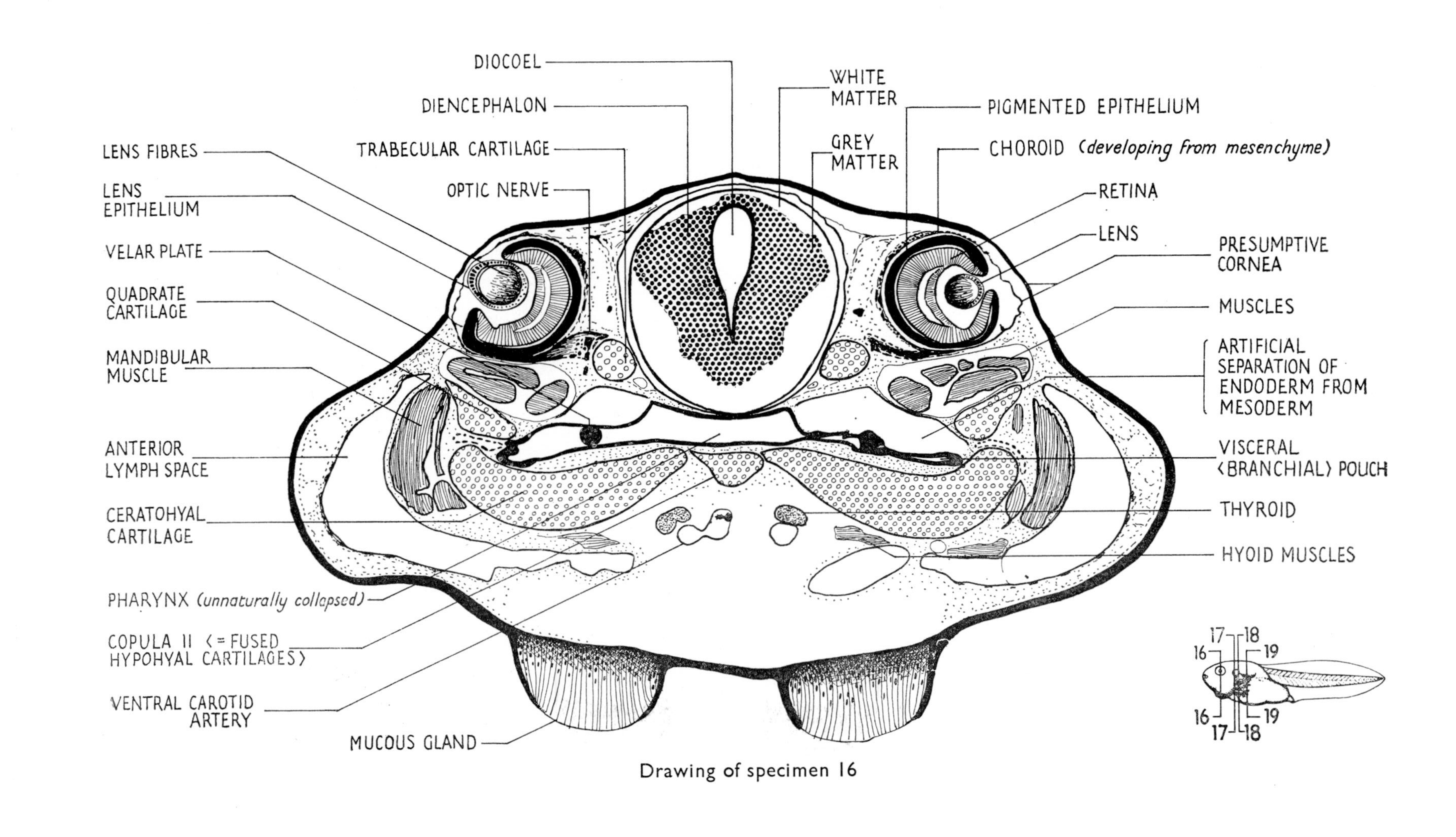
DIOCOEL
DIENCEPHALON
TRABECULAR CARTILAGE
OPTIC NERVE
WHITE MATTER
GREY MATTER
PIGMENTED EPITHELIUM
CHOROID (developing from mesenchyme)
RETINA
LENS
PRESUMPTIVE CORNEA
MUSCLES
ARTIFICIAL SEPARATION OF ENDODERM FROM MESODERM
VISCERAL (BRANCHIAL) POUCH
THYROID
HYOID MUSCLES
LENS FIBRES
LENS EPITHELIUM
VELAR PLATE
QUADRATE CARTILAGE
MANDIBULAR MUSCLE
ANTERIOR LYMPH SPACE
CERATOHYAL CARTILAGE
PHARYNX (unnaturally collapsed)
COPULA II (= FUSED HYPOHYAL CARTILAGES)
VENTRAL CAROTID ARTERY
MUCOUS GLAND
16
17
18
19

Drawing of specimen 16

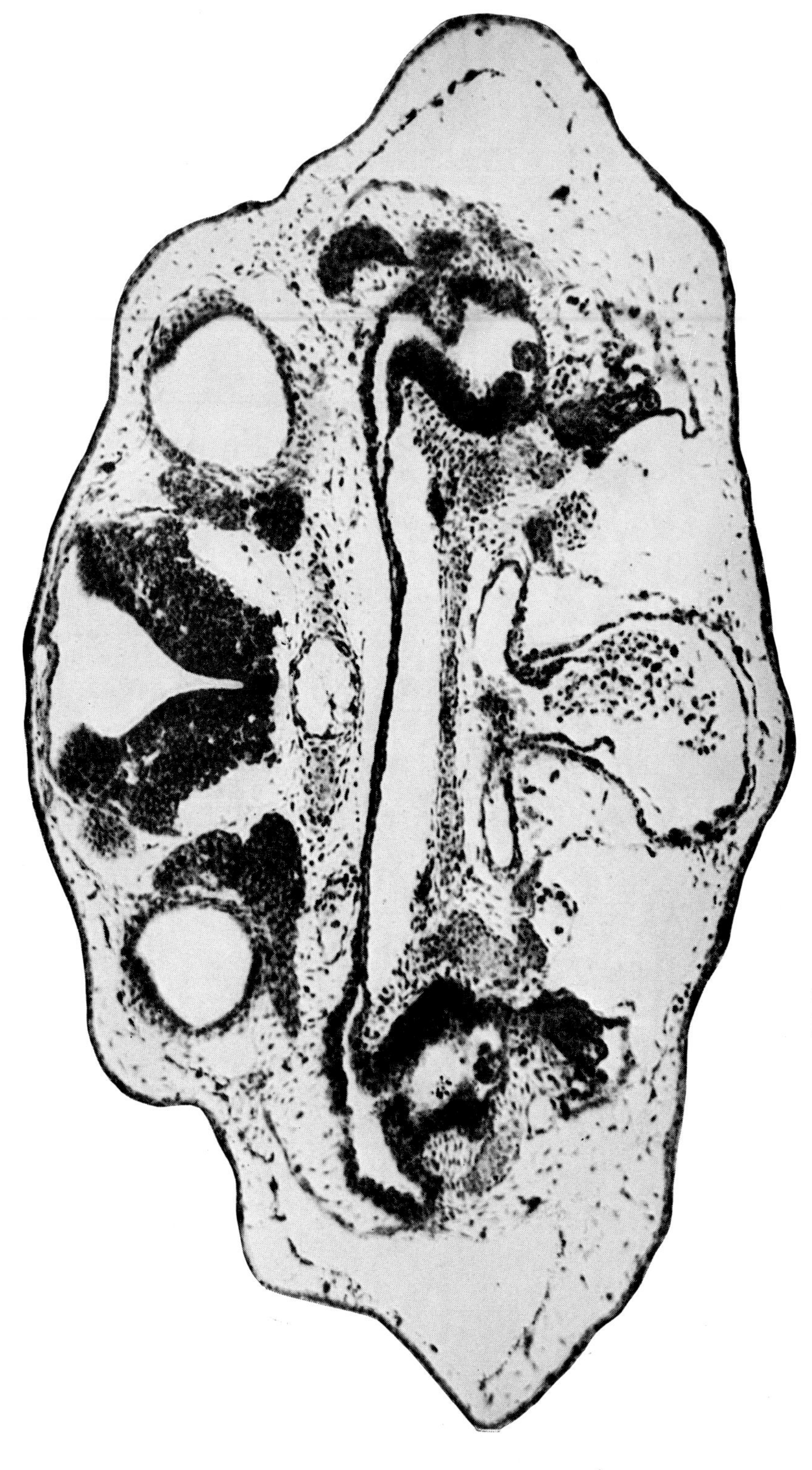

17. **Frog:** external gill larva, auditory region, T.S. *mag. 100×*

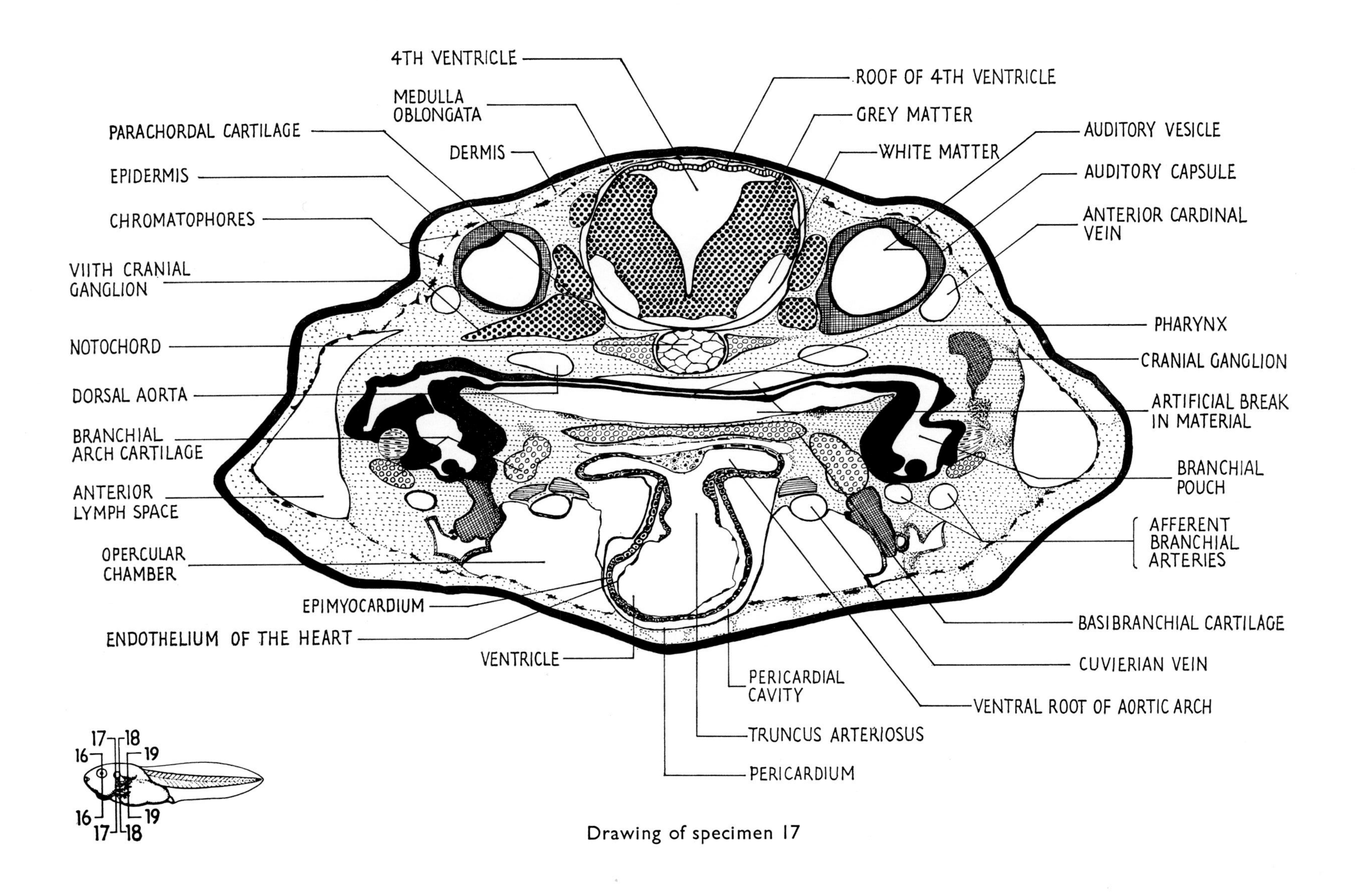
4TH VENTRICLE
MEDULLA OBLONGATA
PARACHORDAL CARTILAGE
DERMIS
EPIDERMIS
CHROMATOPHORES
VIITH CRANIAL GANGLION
NOTOCHORD
DORSAL AORTA
BRANCHIAL ARCH CARTILAGE
ANTERIOR LYMPH SPACE
OPERCULAR CHAMBER
EPIMYOCARDIUM
ENDOTHELIUM OF THE HEART
VENTRICLE
PERICARDIAL CAVITY
TRUNCUS ARTERIOSUS
PERICARDIUM
ROOF OF 4TH VENTRICLE
GREY MATTER
WHITE MATTER
AUDITORY VESICLE
AUDITORY CAPSULE
ANTERIOR CARDINAL VEIN
PHARYNX
CRANIAL GANGLION
ARTIFICIAL BREAK IN MATERIAL
BRANCHIAL POUCH
AFFERENT BRANCHIAL ARTERIES
BASIBRANCHIAL CARTILAGE
CUVIERIAN VEIN
VENTRAL ROOT OF AORTIC ARCH
16
17
18
19
16
17
18
19

Drawing of specimen 17

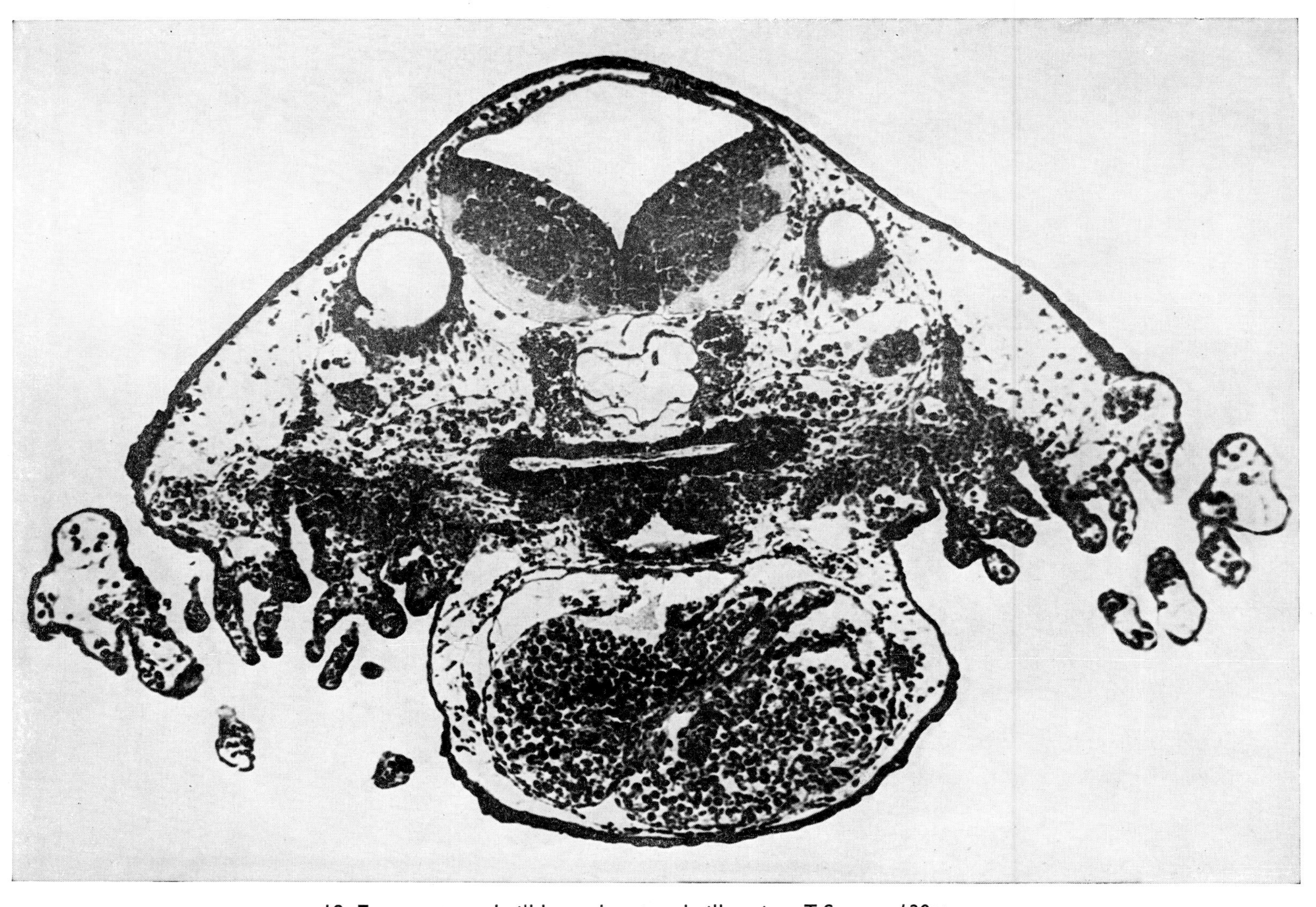

18. **Frog:** external gill larva, heart and gill region, T.S. *mag. 120×*

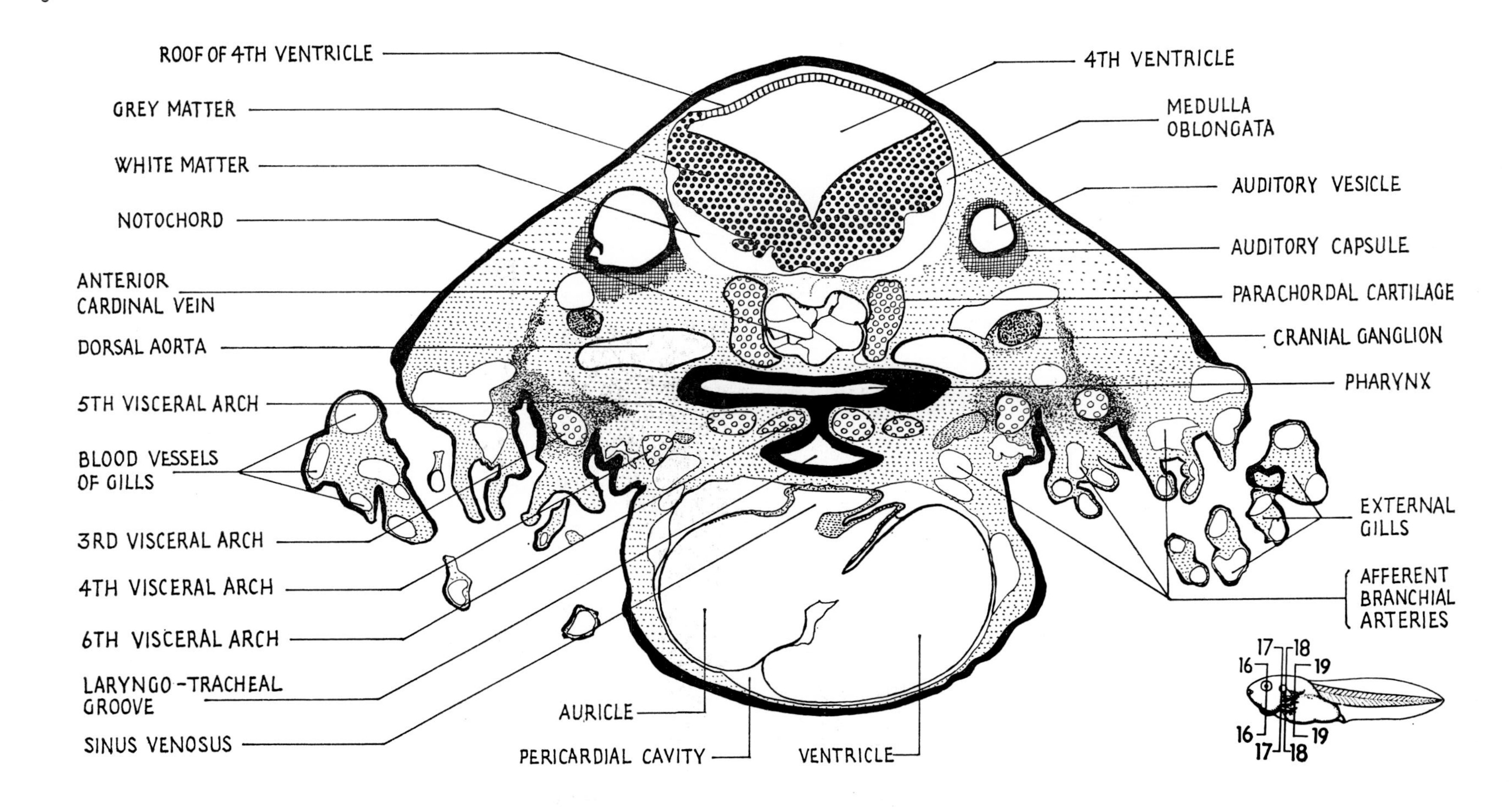
ROOF OF 4TH VENTRICLE
GREY MATTER
WHITE MATTER
NOTOCHORD
ANTERIOR CARDINAL VEIN
DORSAL AORTA
5TH VISCERAL ARCH
BLOOD VESSELS OF GILLS
3RD VISCERAL ARCH
4TH VISCERAL ARCH
6TH VISCERAL ARCH
LARYNGO-TRACHEAL GROOVE
SINUS VENOSUS
AURICLE
PERICARDIAL CAVITY
VENTRICLE
4TH VENTRICLE
MEDULLA OBLONGATA
AUDITORY VESICLE
AUDITORY CAPSULE
PARACHORDAL CARTILAGE
CRANIAL GANGLION
PHARYNX
EXTERNAL GILLS
AFFERENT BRANCHIAL ARTERIES
16
17
18
19
16
17
18
19

Drawing of specimen 18

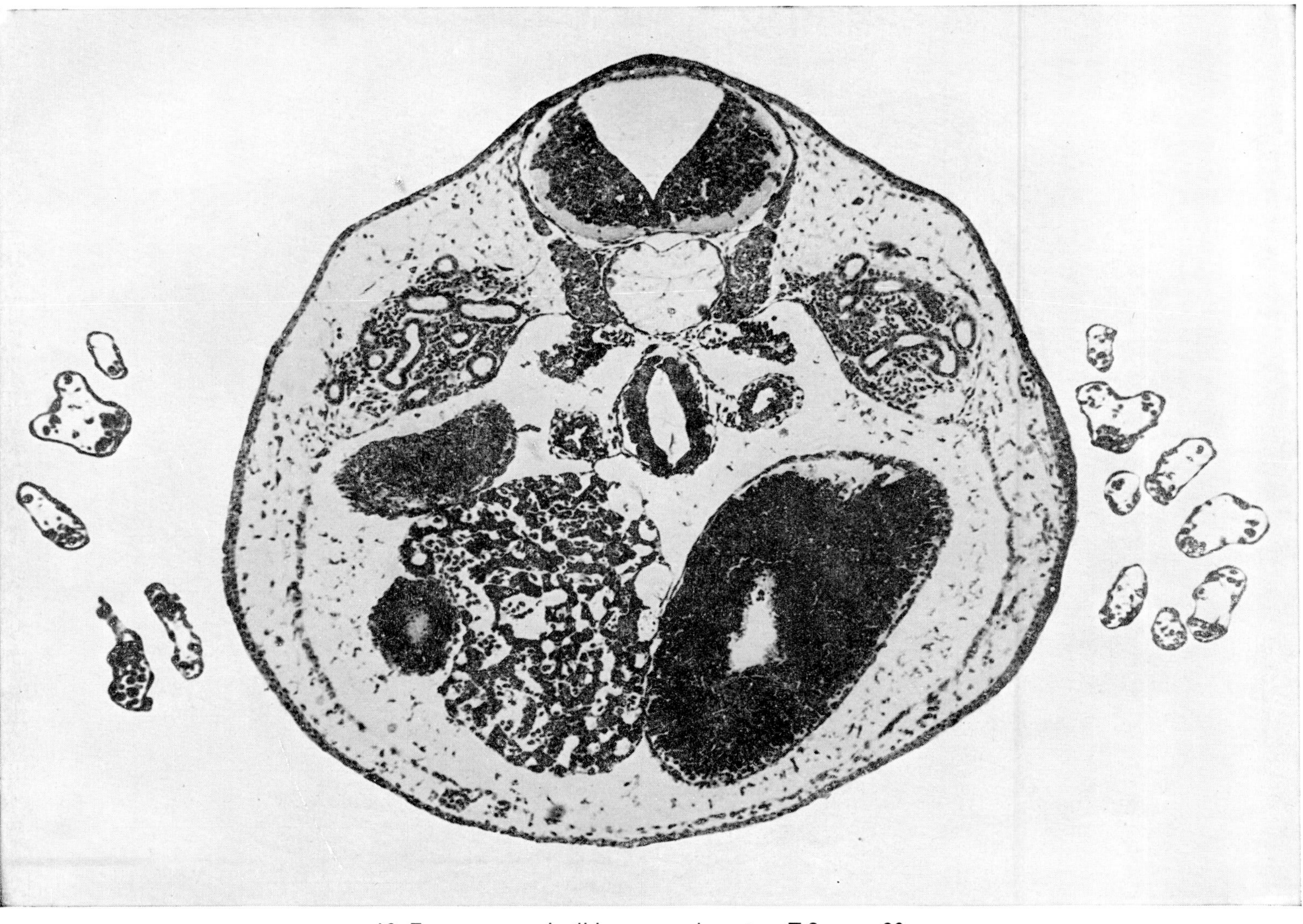

19. **Frog:** external gill larva, trunk region, T.S. *mag. 80*×

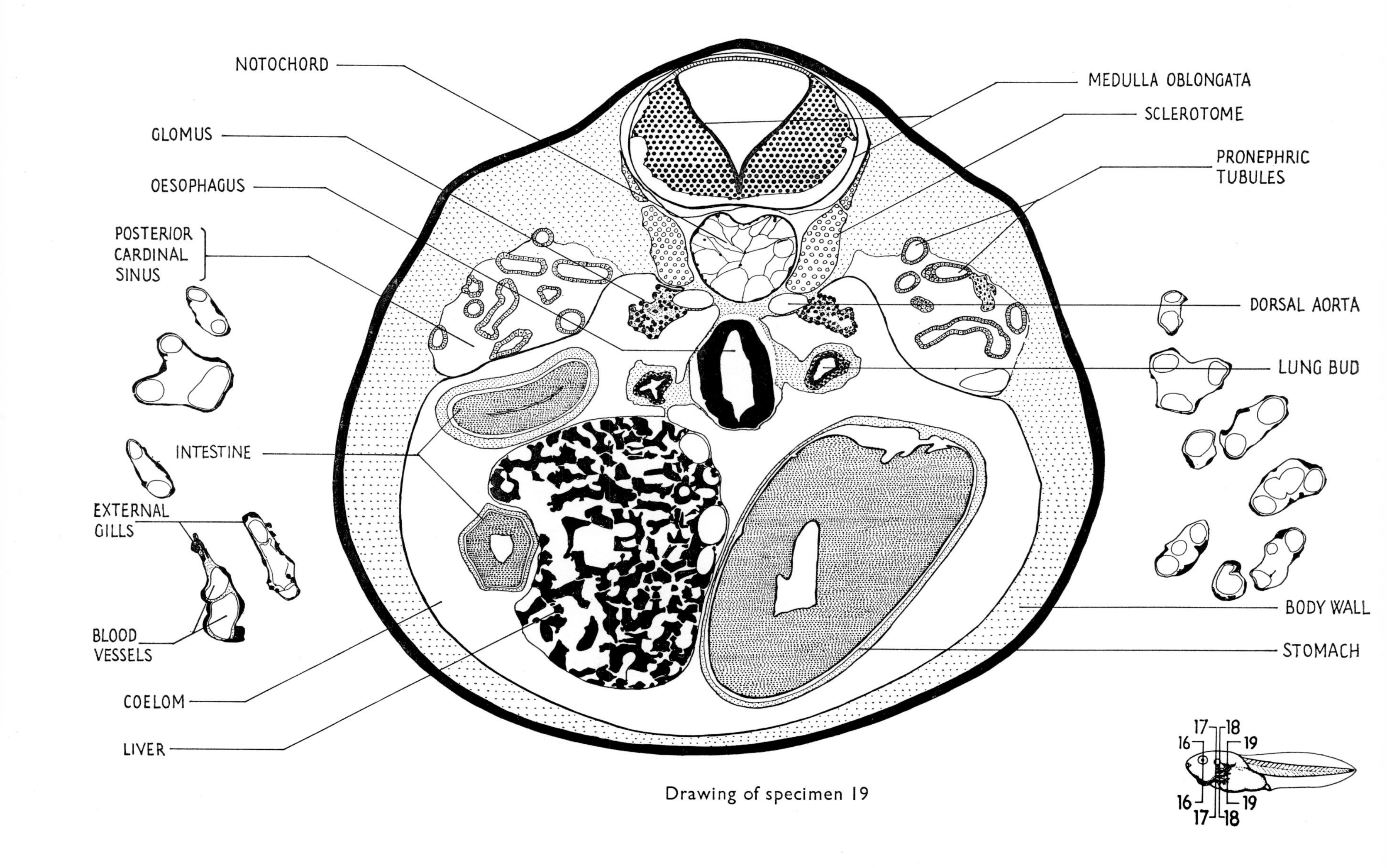
NOTOCHORD
GLOMUS
OESOPHAGUS
POSTERIOR CARDINAL SINUS
INTESTINE
EXTERNAL GILLS
BLOOD VESSELS
COELOM
LIVER
MEDULLA OBLONGATA
SCLEROTOME
PRONEPHRIC TUBULES
DORSAL AORTA
LUNG BUD
BODY WALL
STOMACH
16
17
18
19
16
17
18
19

Drawing of specimen 19

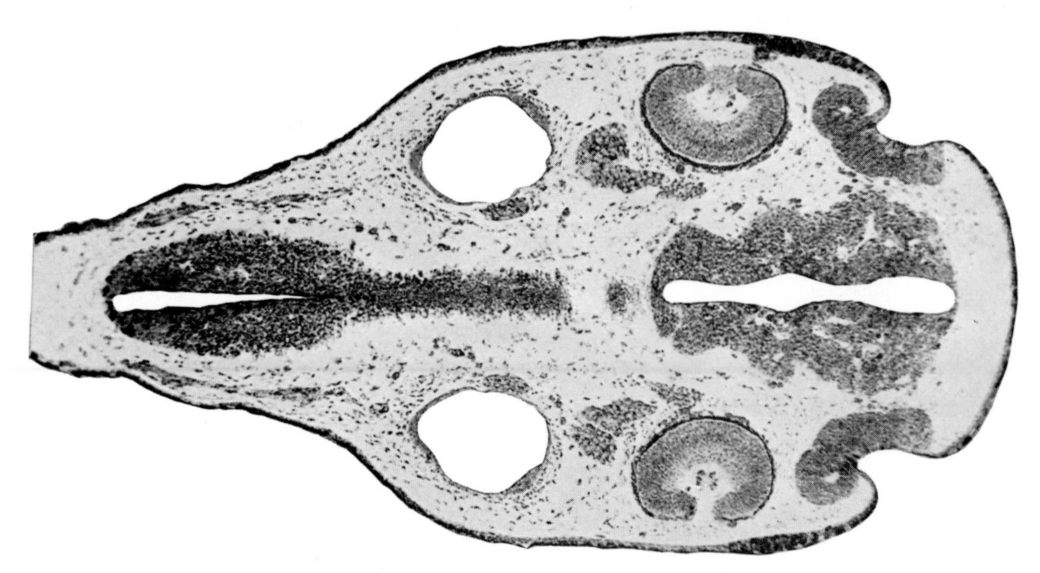

20. **Frog:** external gill larva, head region, H.L.S. *mag. 85×*

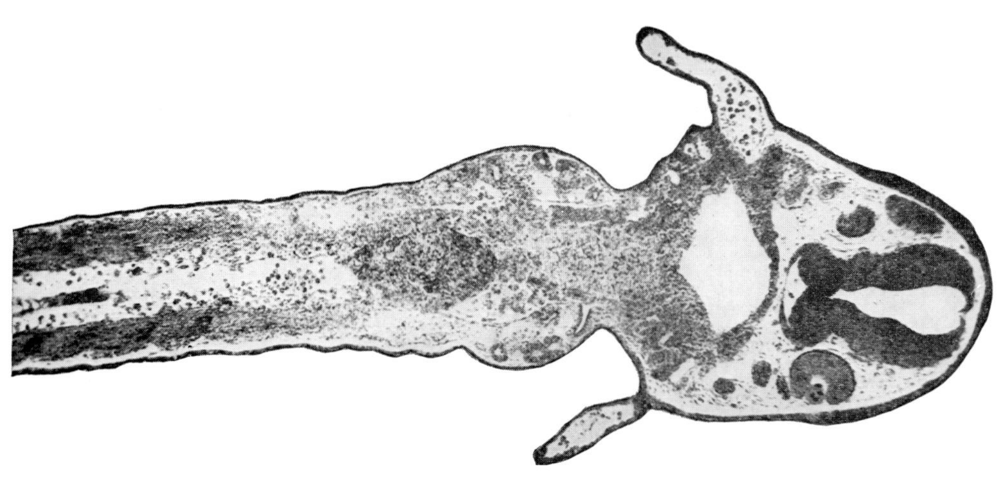

21. **Frog:** external gill larva, trunk region, H.L.S. *mag. 50×*

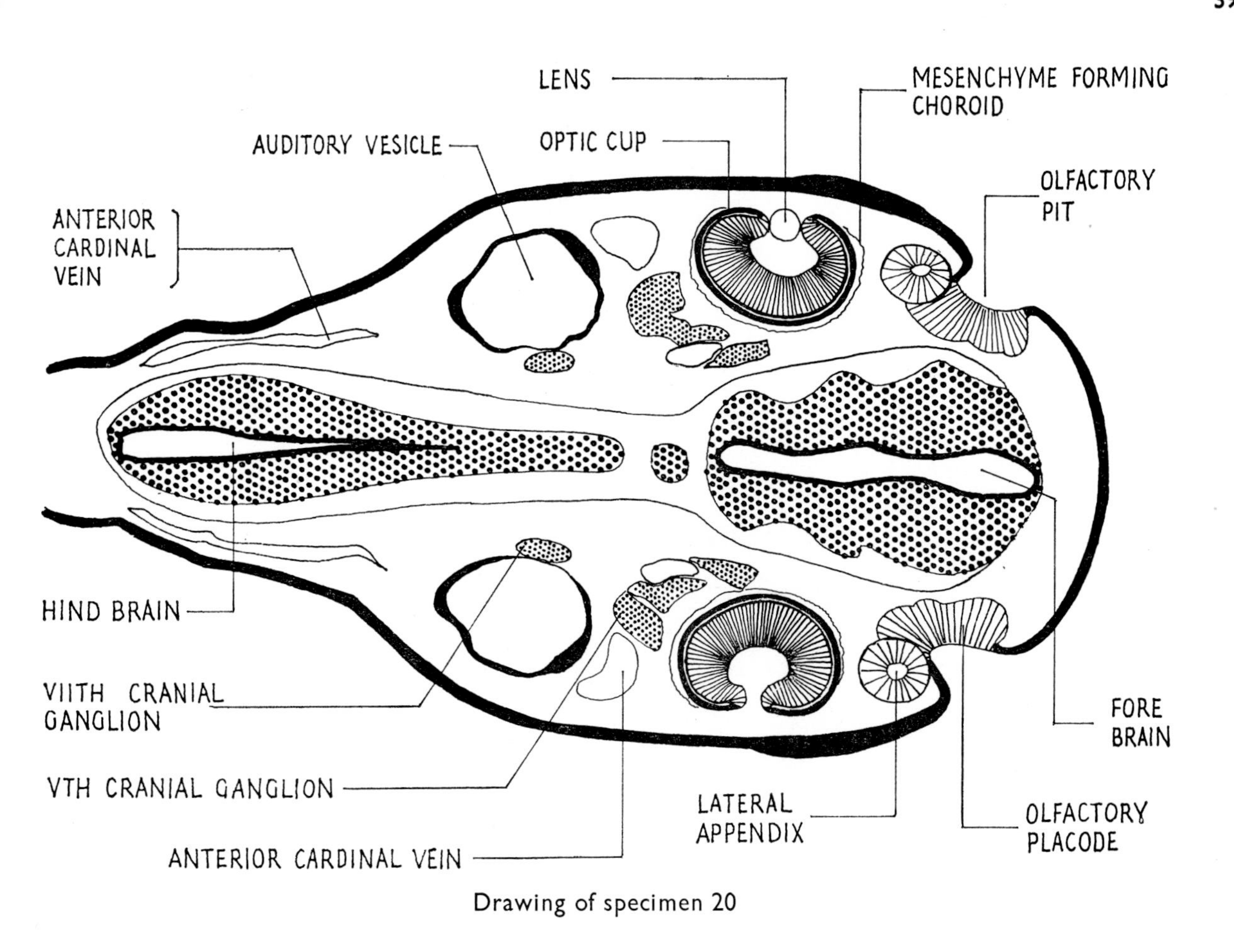
LENS
MESENCHYME FORMING CHOROID
AUDITORY VESICLE
OPTIC CUP
OLFACTORY PIT
ANTERIOR CARDINAL VEIN
HIND BRAIN
VIITH CRANIAL GANGLION
FORE BRAIN
VTH CRANIAL GANGLION
LATERAL APPENDIX
OLFACTORY PLACODE
ANTERIOR CARDINAL VEIN

Drawing of specimen 20

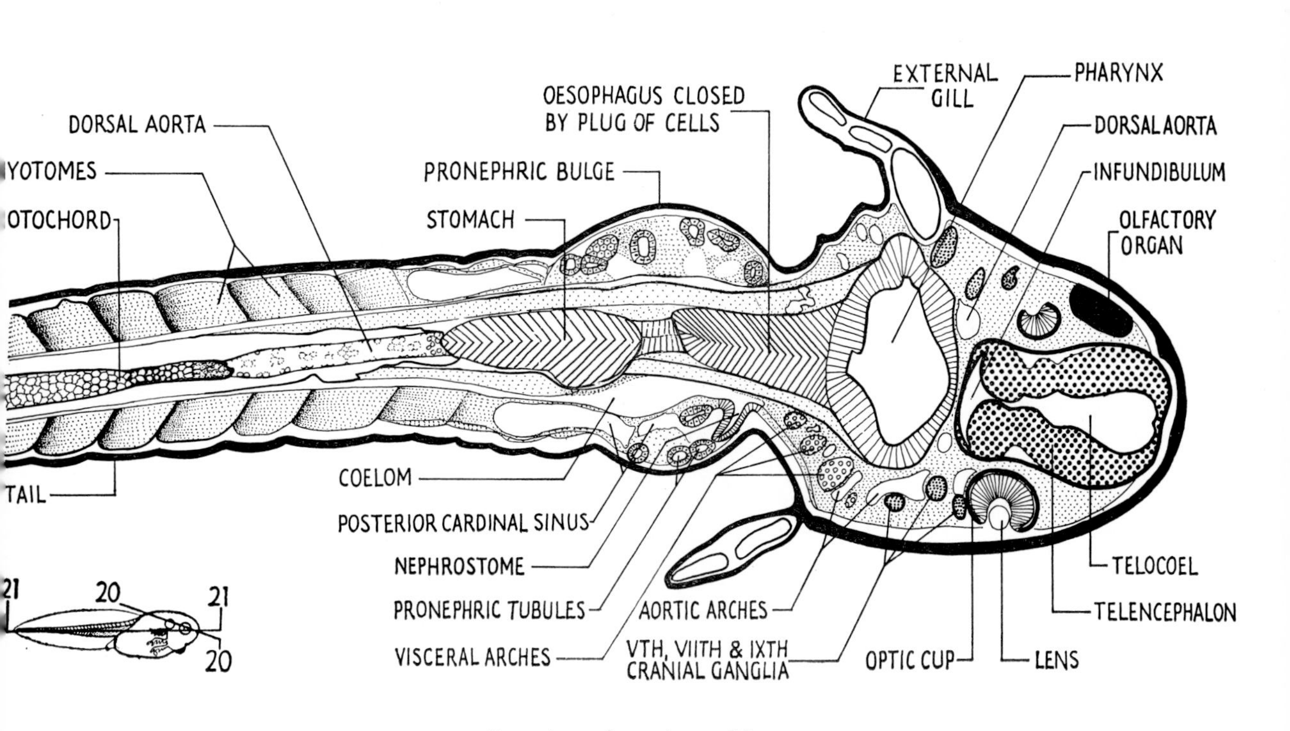
EXTERNAL GILL
PHARYNX
OESOPHAGUS CLOSED BY PLUG OF CELLS
DORSAL AORTA
DORSAL AORTA
YOTOMES
INFUNDIBULUM
PRONEPHRIC BULGE
OTOCHORD
STOMACH
OLFACTORY ORGAN
TAIL
COELOM
POSTERIOR CARDINAL SINUS
NEPHROSTOME
TELOCOEL
21
20
21
20
PRONEPHRIC TUBULES
AORTIC ARCHES
TELENCEPHALON
VISCERAL ARCHES
VTH, VIITH & IXTH CRANIAL GANGLIA
OPTIC CUP
LENS

Drawing of specimen 21

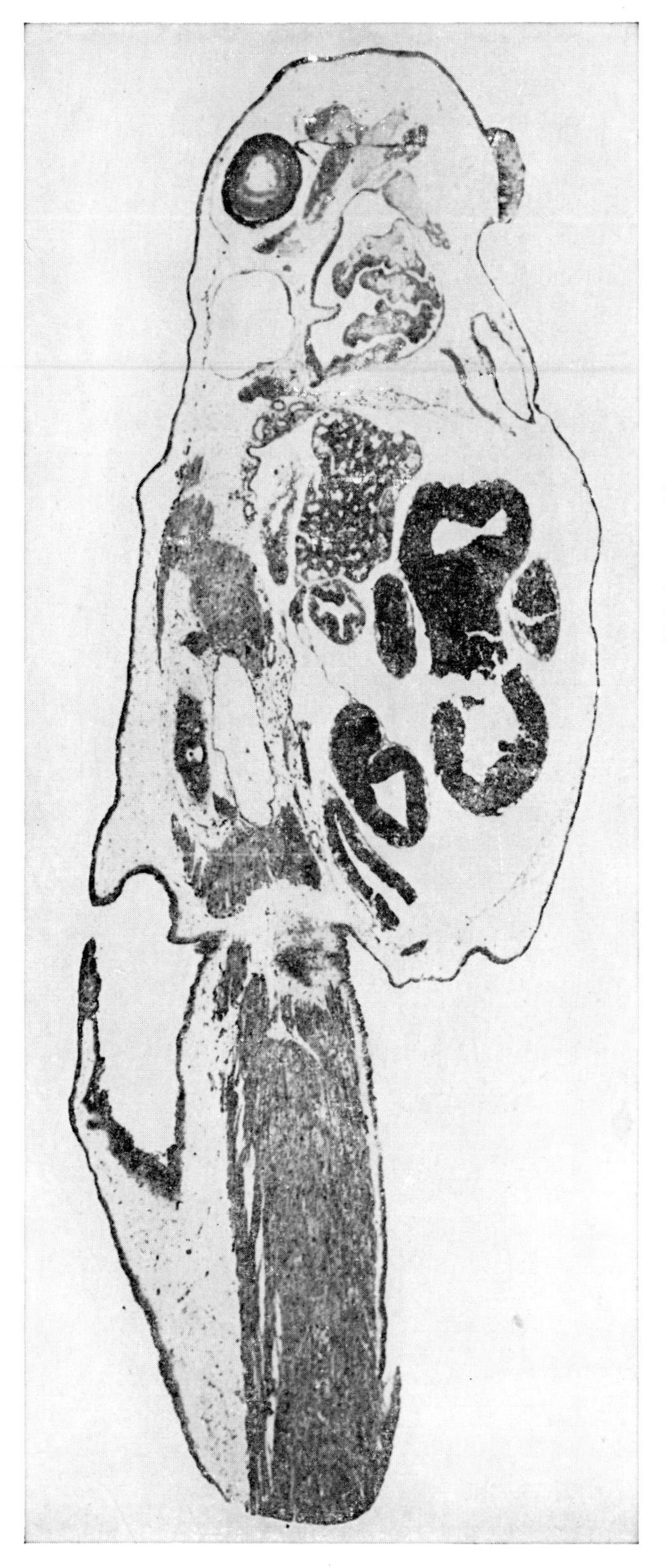

22. **Frog:** internal gill larva, trunk region, V.L.S. *mag. 40×*

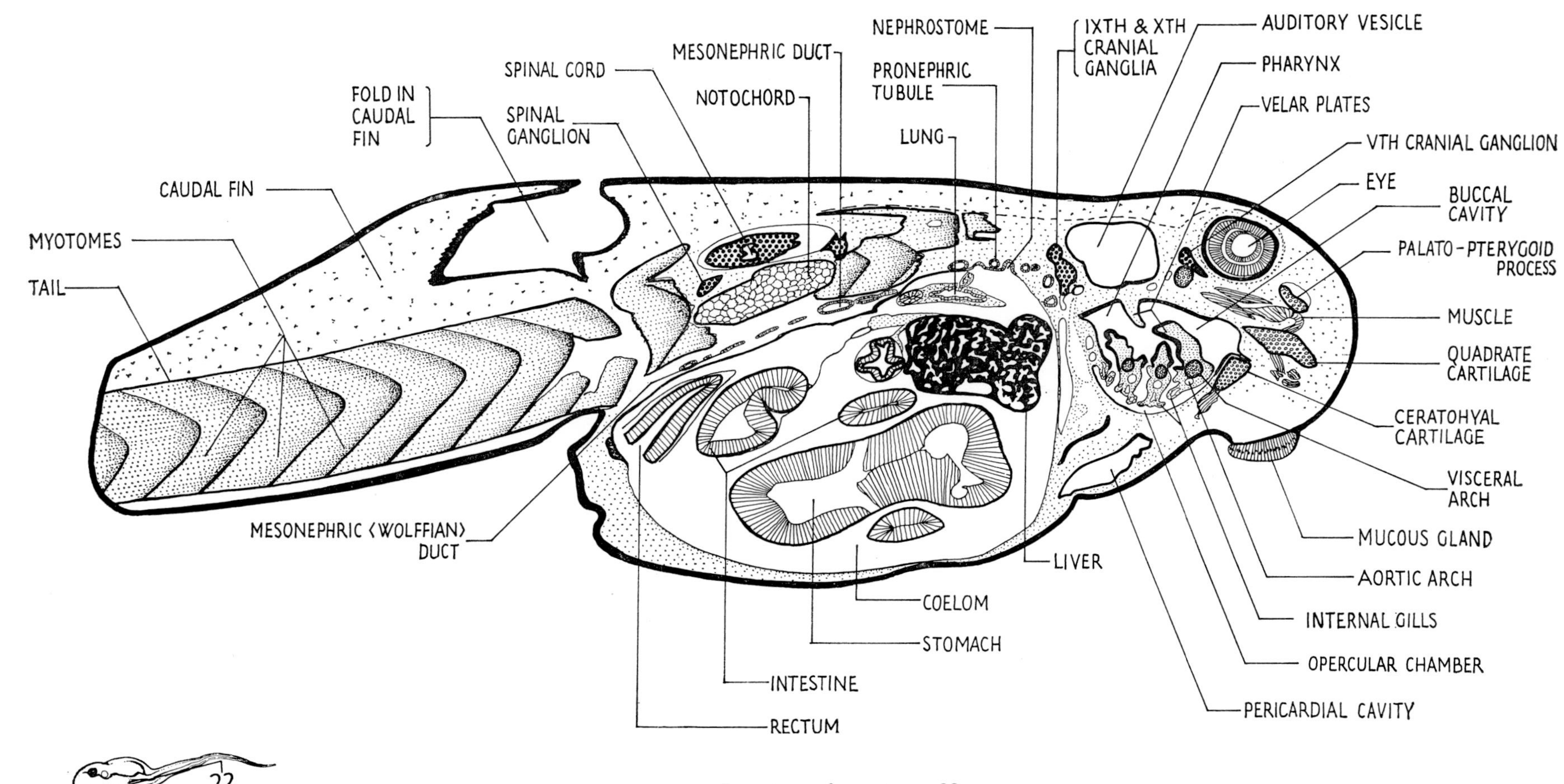

FOLD IN CAUDAL FIN
SPINAL CORD
SPINAL GANGLION
MESONEPHRIC DUCT
NOTOCHORD
NEPHROSTOME
PRONEPHRIC TUBULE
LUNG
IXTH & XTH CRANIAL GANGLIA
AUDITORY VESICLE
PHARYNX
VELAR PLATES
VTH CRANIAL GANGLION
EYE
BUCCAL CAVITY
PALATO-PTERYGOID PROCESS
MUSCLE
QUADRATE CARTILAGE
CERATOHYAL CARTILAGE
VISCERAL ARCH
MUCOUS GLAND
AORTIC ARCH
INTERNAL GILLS
OPERCULAR CHAMBER
PERICARDIAL CAVITY
CAUDAL FIN
MYOTOMES
TAIL
MESONEPHRIC ‹WOLFFIAN› DUCT
LIVER
COELOM
STOMACH
INTESTINE
RECTUM
22
22

Drawing of specimen 22

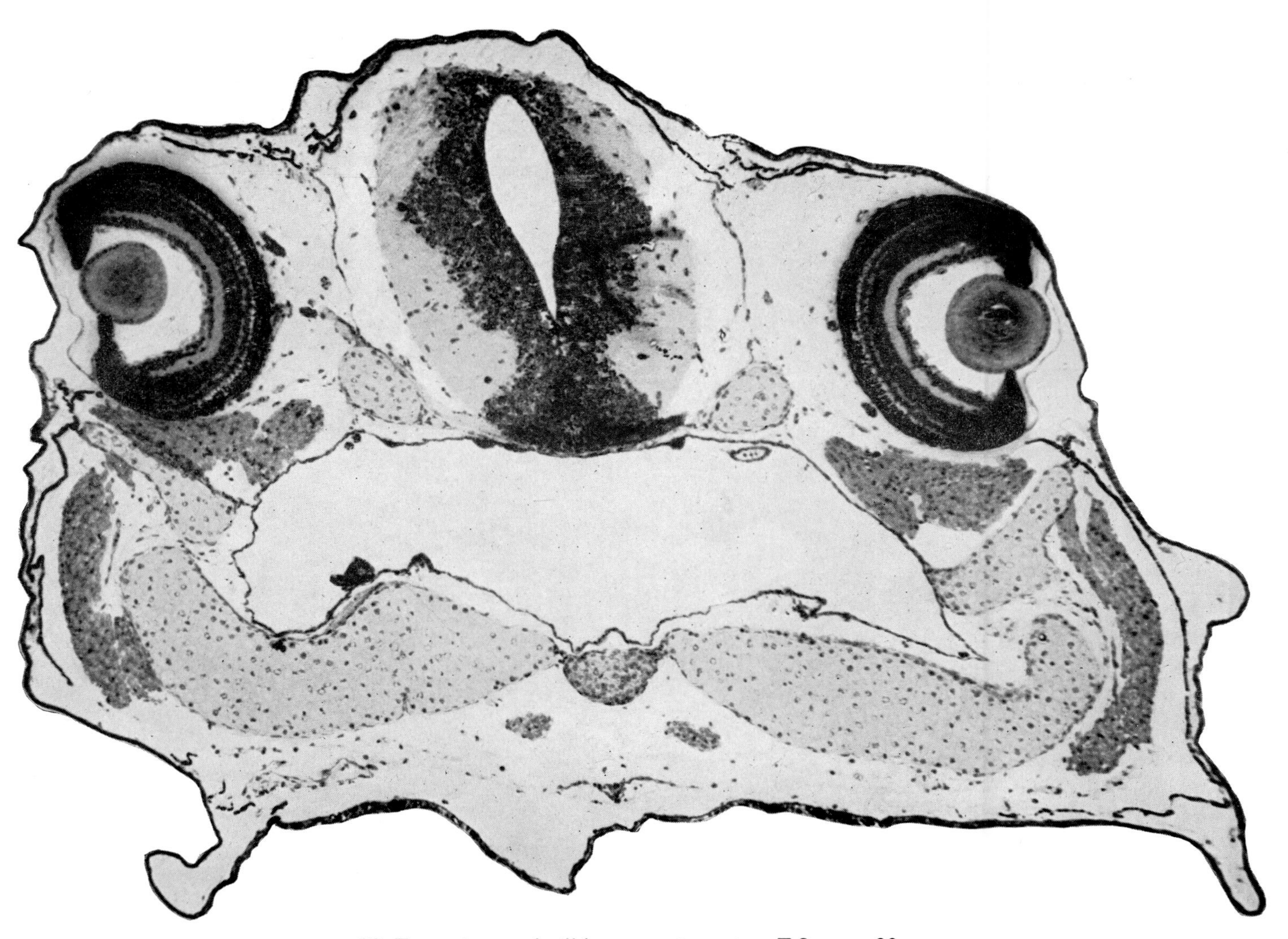

23. **Frog:** internal gill larva, optic region, T.S. *mag. 80×*

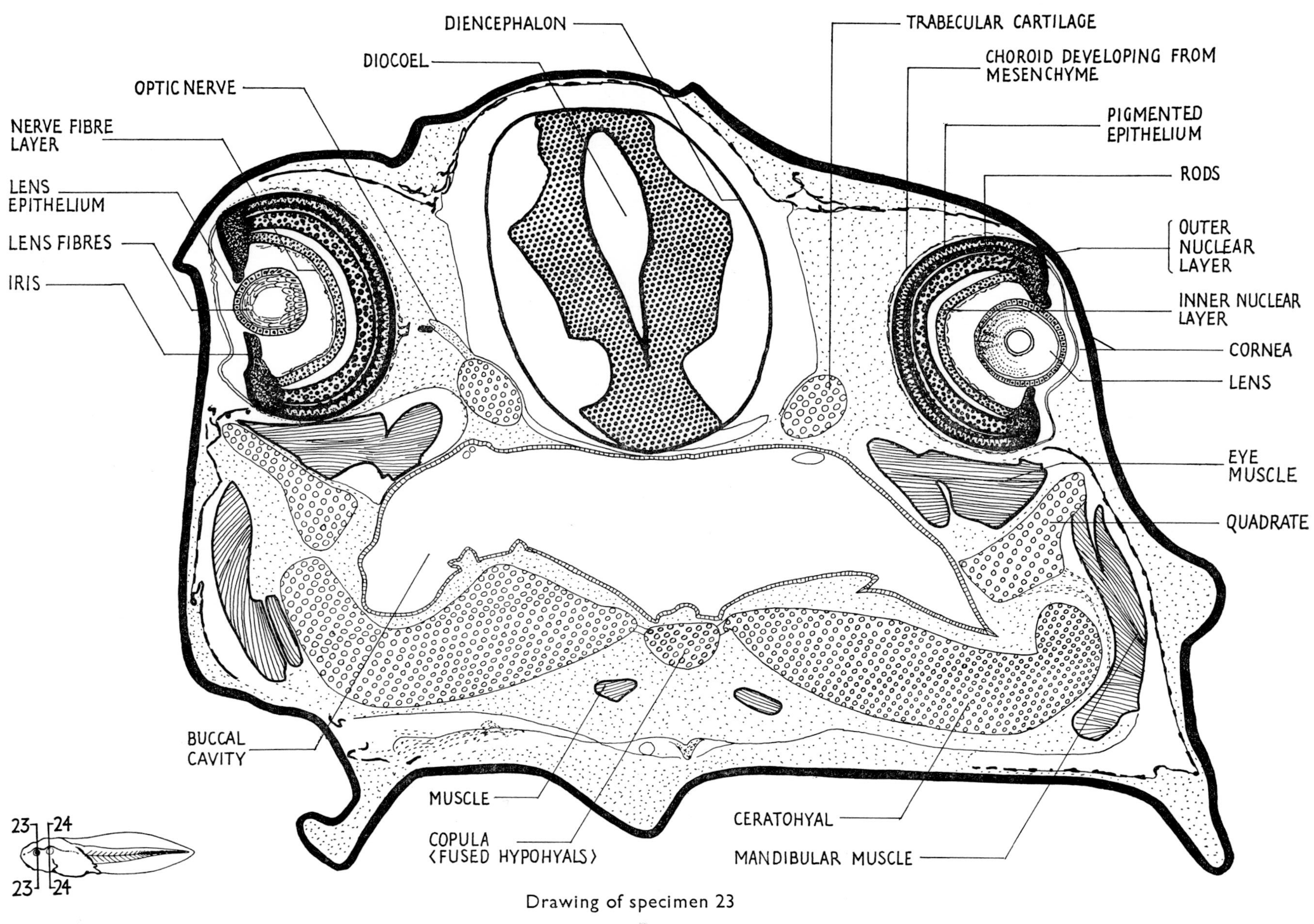
DIENCEPHALON
DIOCOEL
OPTIC NERVE
NERVE FIBRE LAYER
LENS EPITHELIUM
LENS FIBRES
IRIS
TRABECULAR CARTILAGE
CHOROID DEVELOPING FROM MESENCHYME
PIGMENTED EPITHELIUM
RODS
OUTER NUCLEAR LAYER
INNER NUCLEAR LAYER
CORNEA
LENS
EYE MUSCLE
QUADRATE
BUCCAL CAVITY
MUSCLE
COPULA
<FUSED HYPOHYALS>
CERATOHYAL
MANDIBULAR MUSCLE
23
24
23
24

Drawing of specimen 23

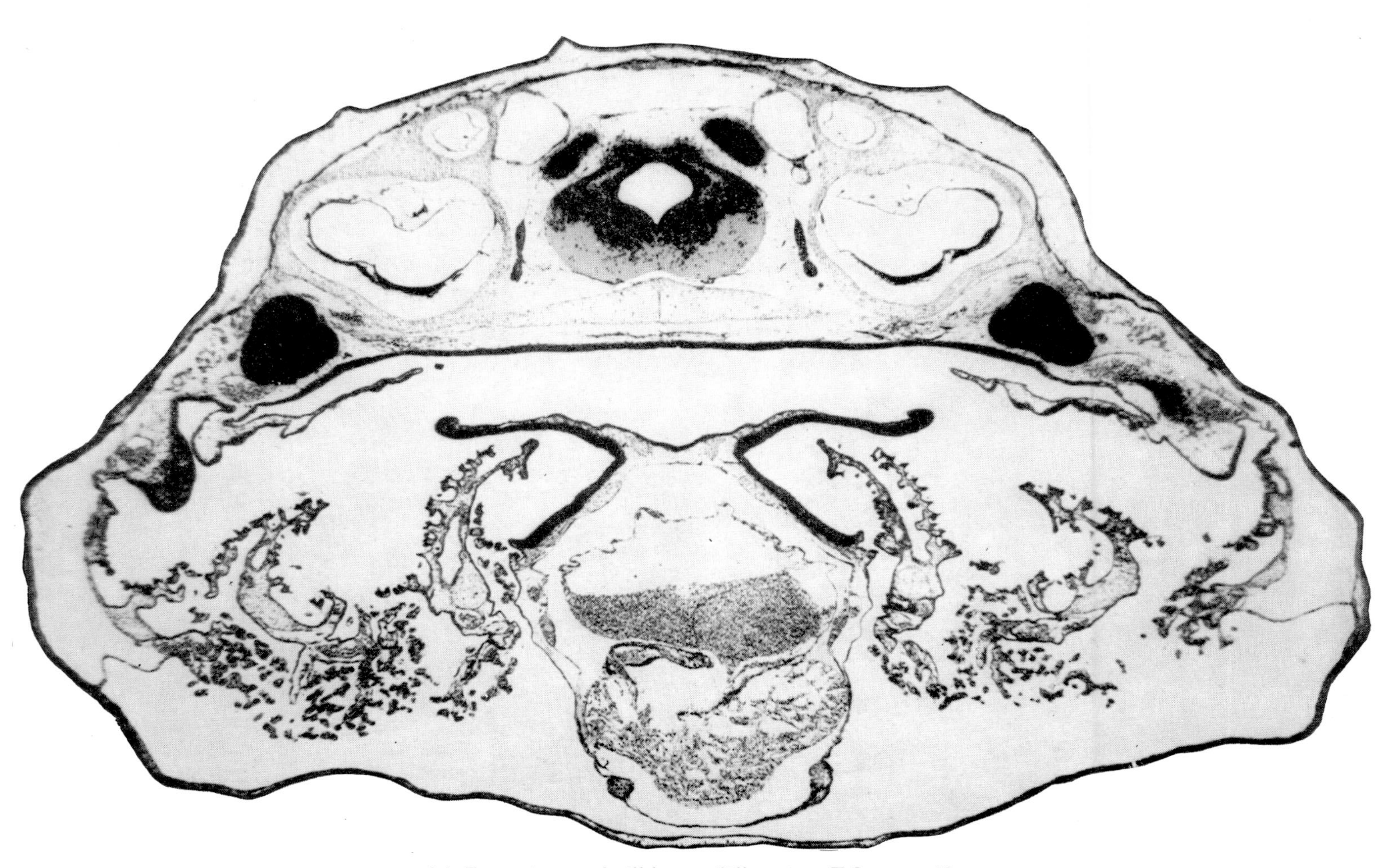

24. **Frog:** internal gill larva, Gill region, T.S. *mag.* 45×

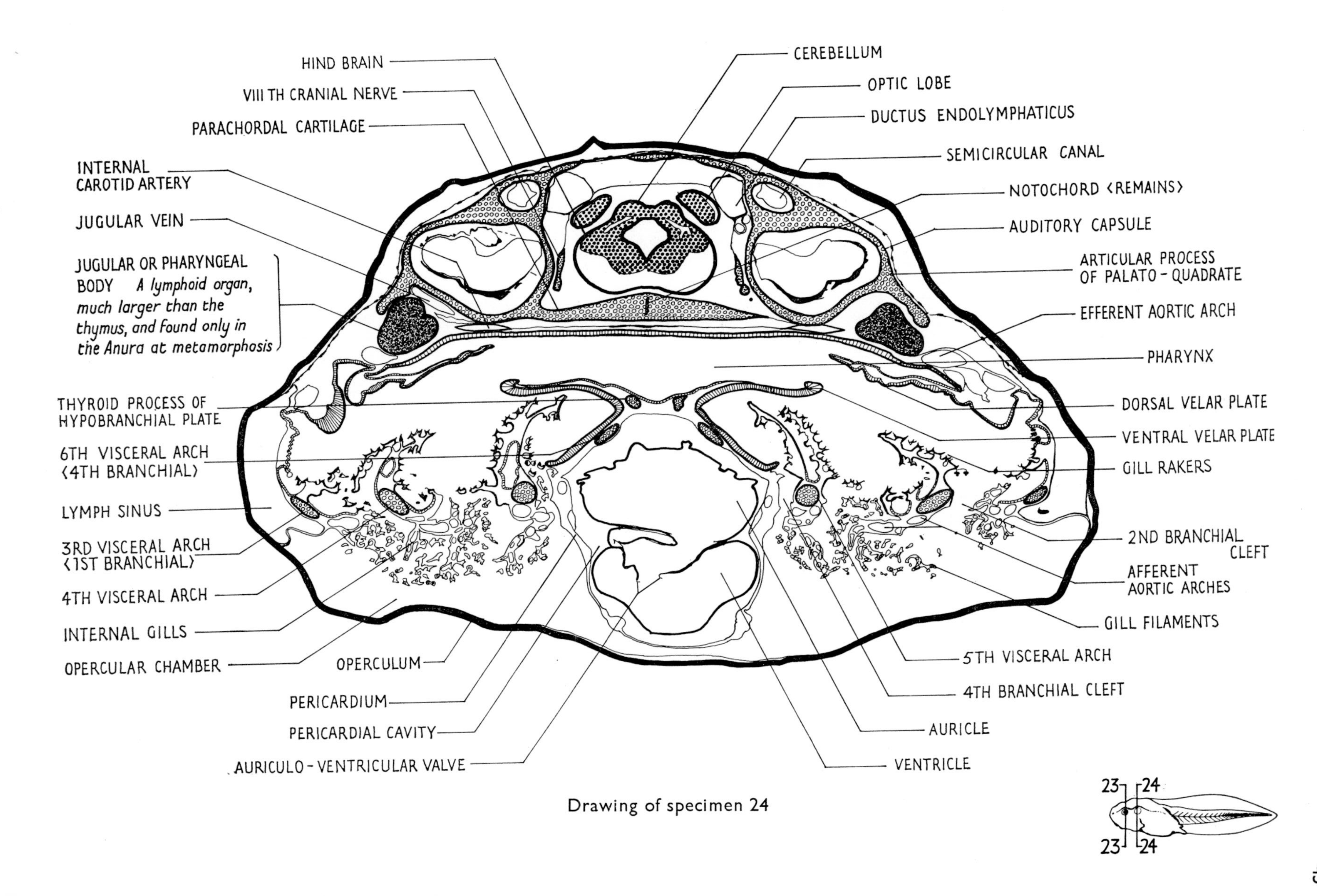
HIND BRAIN
VIII TH CRANIAL NERVE
PARACHORDAL CARTILAGE
INTERNAL CAROTID ARTERY
JUGULAR VEIN
JUGULAR OR PHARYNGEAL BODY *A lymphoid organ, much larger than the thymus, and found only in the Anura at metamorphosis*
THYROID PROCESS OF HYPOBRANCHIAL PLATE
6TH VISCERAL ARCH ‹4TH BRANCHIAL›
LYMPH SINUS
3RD VISCERAL ARCH ‹1ST BRANCHIAL›
4TH VISCERAL ARCH
INTERNAL GILLS
OPERCULAR CHAMBER
OPERCULUM
PERICARDIUM
PERICARDIAL CAVITY
AURICULO-VENTRICULAR VALVE
CEREBELLUM
OPTIC LOBE
DUCTUS ENDOLYMPHATICUS
SEMICIRCULAR CANAL
NOTOCHORD ‹REMAINS›
AUDITORY CAPSULE
ARTICULAR PROCESS OF PALATO-QUADRATE
EFFERENT AORTIC ARCH
PHARYNX
DORSAL VELAR PLATE
VENTRAL VELAR PLATE
GILL RAKERS
2ND BRANCHIAL CLEFT
AFFERENT AORTIC ARCHES
GILL FILAMENTS
5TH VISCERAL ARCH
4TH BRANCHIAL CLEFT
AURICLE
VENTRICLE
23
24
23
24

Drawing of specimen 24

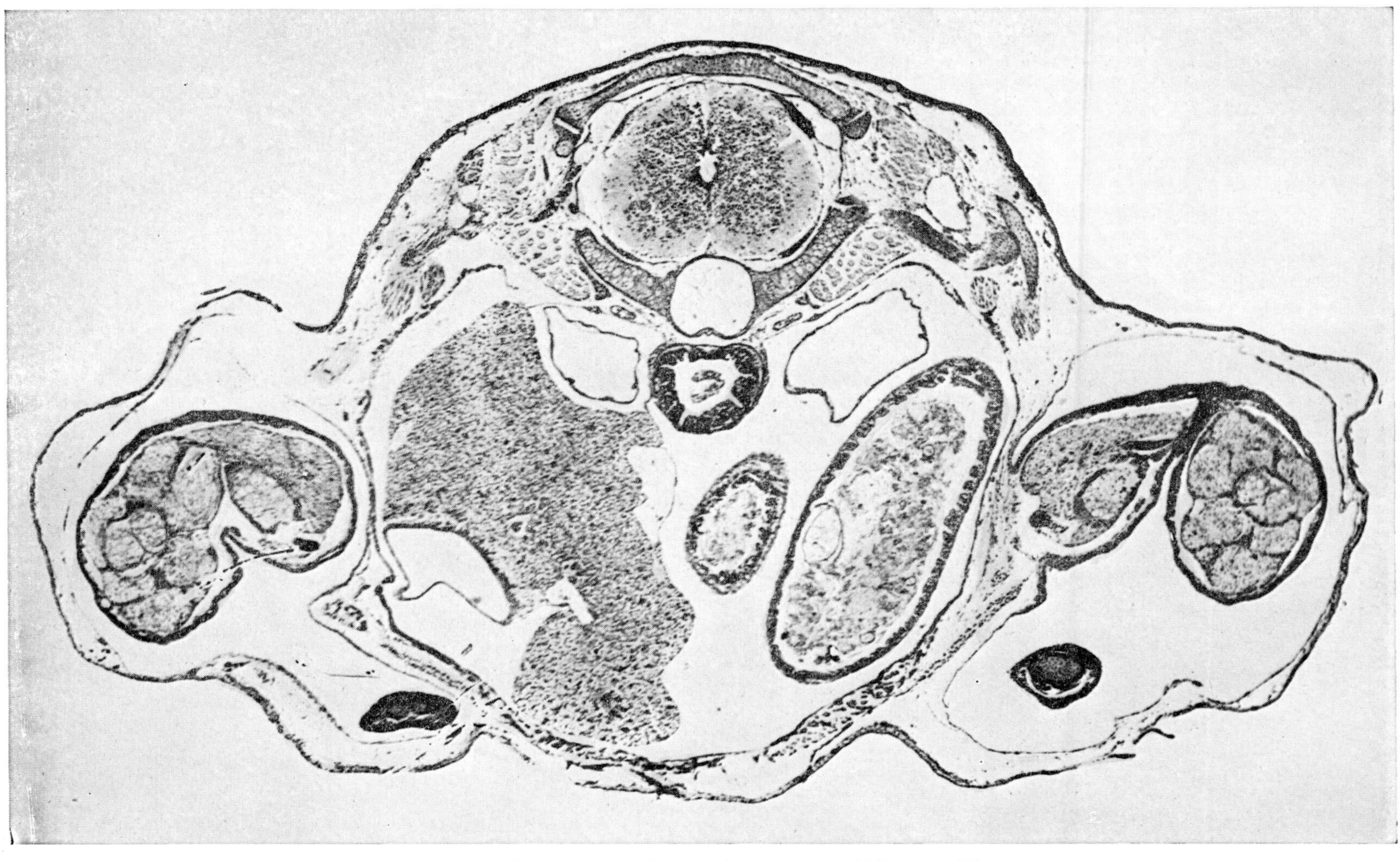

25. **Frog:** 19-mm. tadpole, forelimb region, T.S. *mag.* 35×

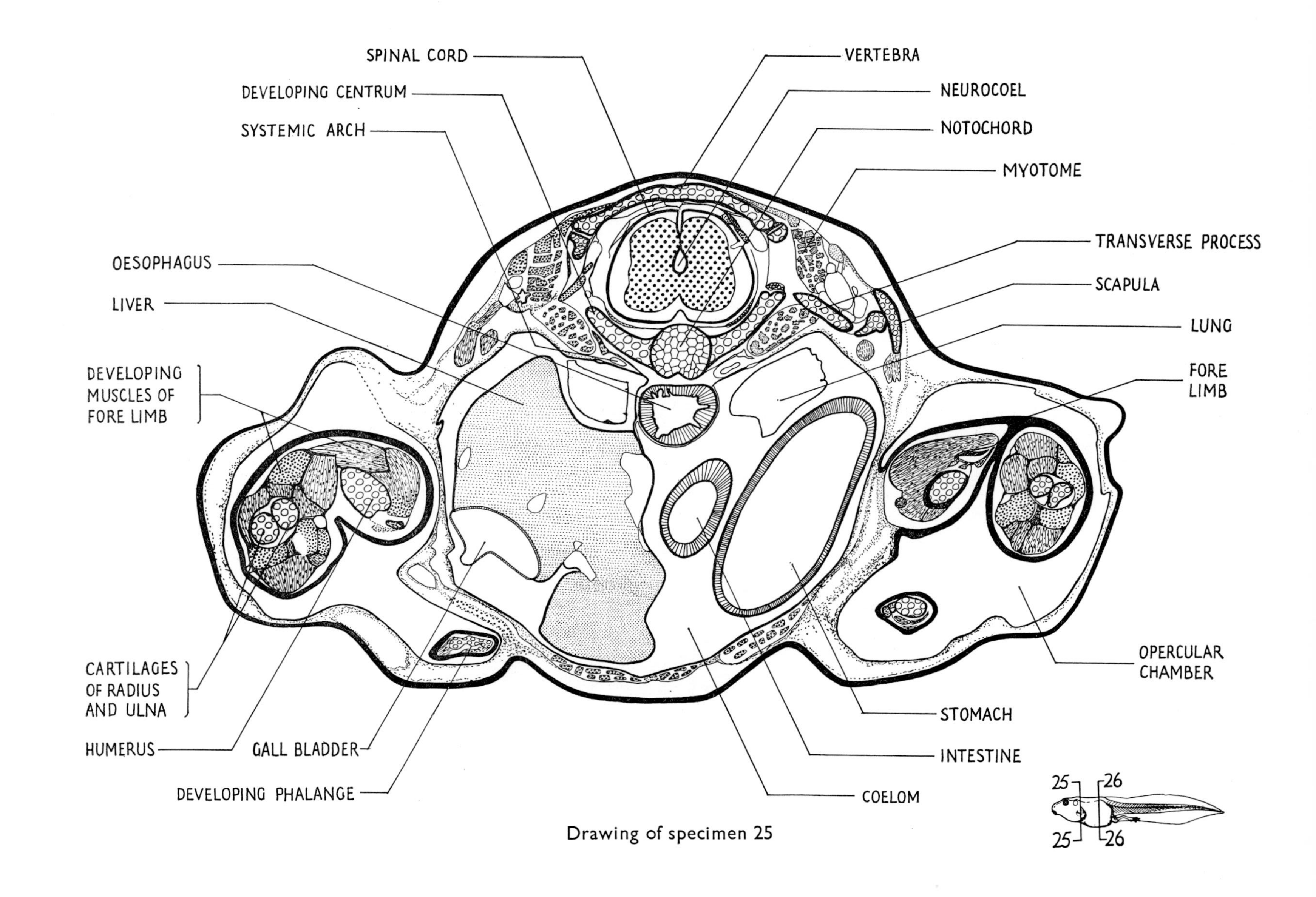

SPINAL CORD
DEVELOPING CENTRUM
SYSTEMIC ARCH
VERTEBRA
NEUROCOEL
NOTOCHORD
MYOTOME
TRANSVERSE PROCESS
SCAPULA
LUNG
FORE LIMB
OESOPHAGUS
LIVER
DEVELOPING MUSCLES OF FORE LIMB
CARTILAGES OF RADIUS AND ULNA
HUMERUS
GALL BLADDER
DEVELOPING PHALANGE
OPERCULAR CHAMBER
STOMACH
INTESTINE
COELOM
25
26
25
26

Drawing of specimen 25

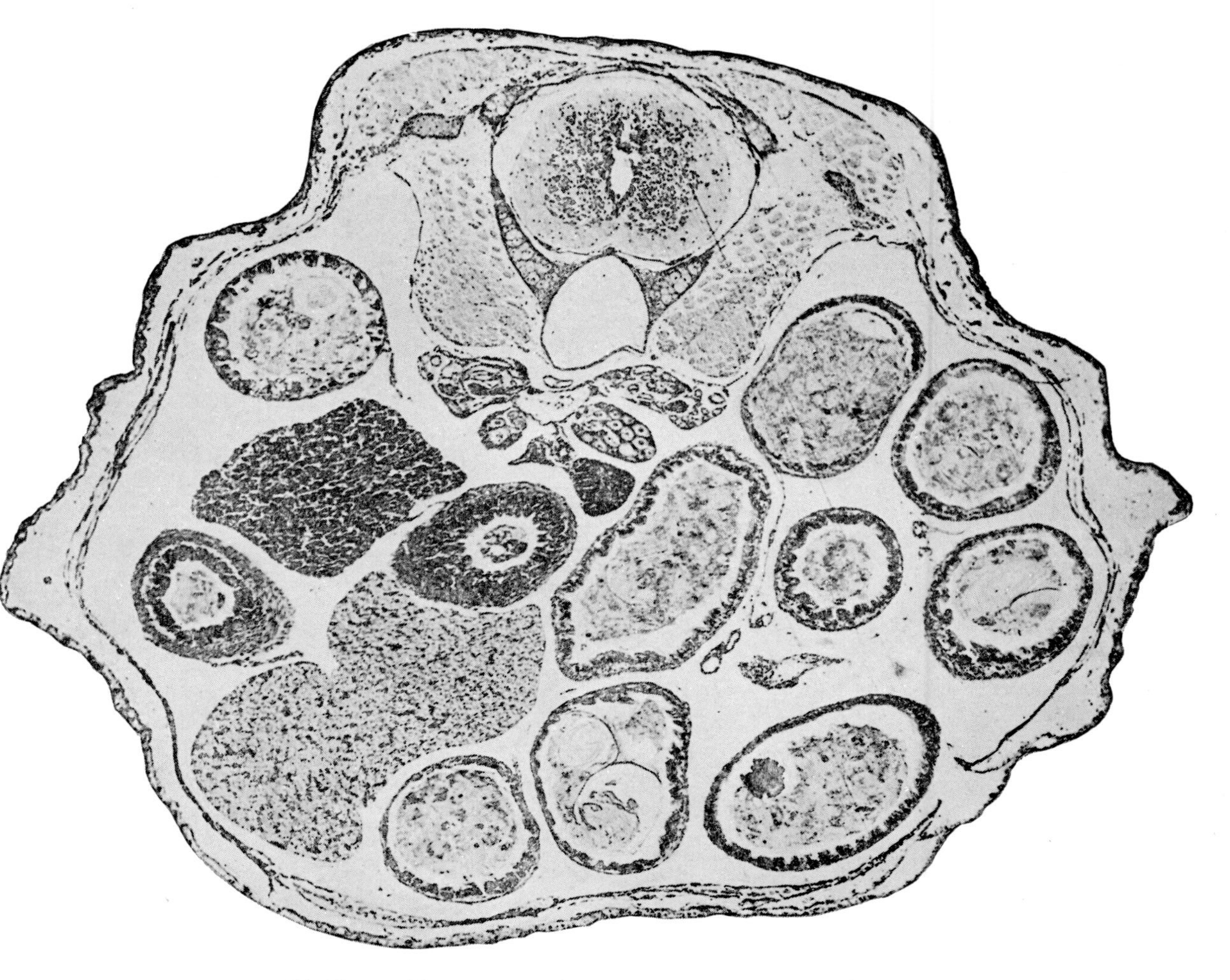

26. **Frog:** 19-mm. tadpole, trunk region, T.S. *mag.* 40×

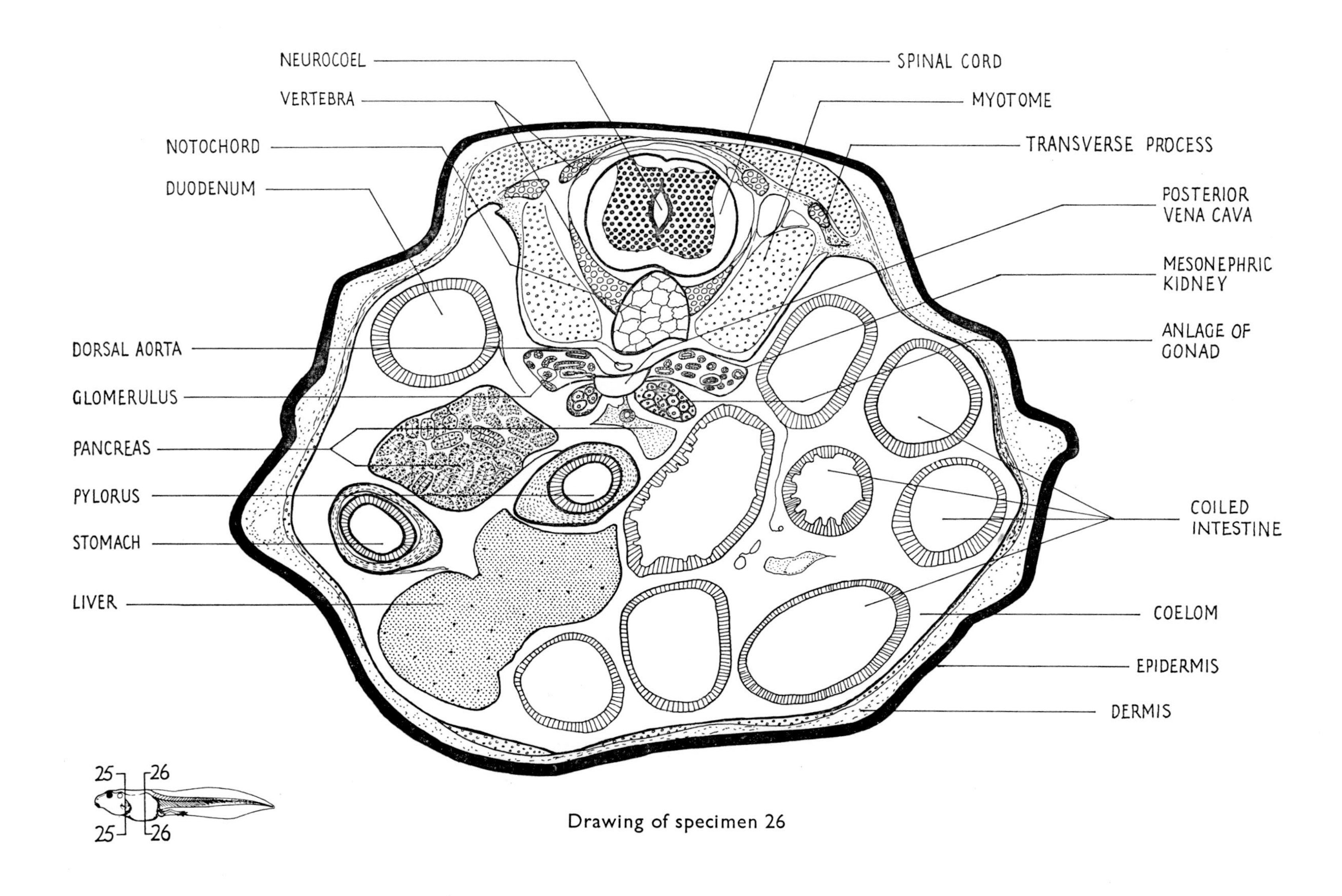
NEUROCOEL
VERTEBRA
NOTOCHORD
DUODENUM
DORSAL AORTA
GLOMERULUS
PANCREAS
PYLORUS
STOMACH
LIVER
SPINAL CORD
MYOTOME
TRANSVERSE PROCESS
POSTERIOR VENA CAVA
MESONEPHRIC KIDNEY
ANLAGE OF GONAD
COILED INTESTINE
COELOM
EPIDERMIS
DERMIS
25
26
25
26

Drawing of specimen 26

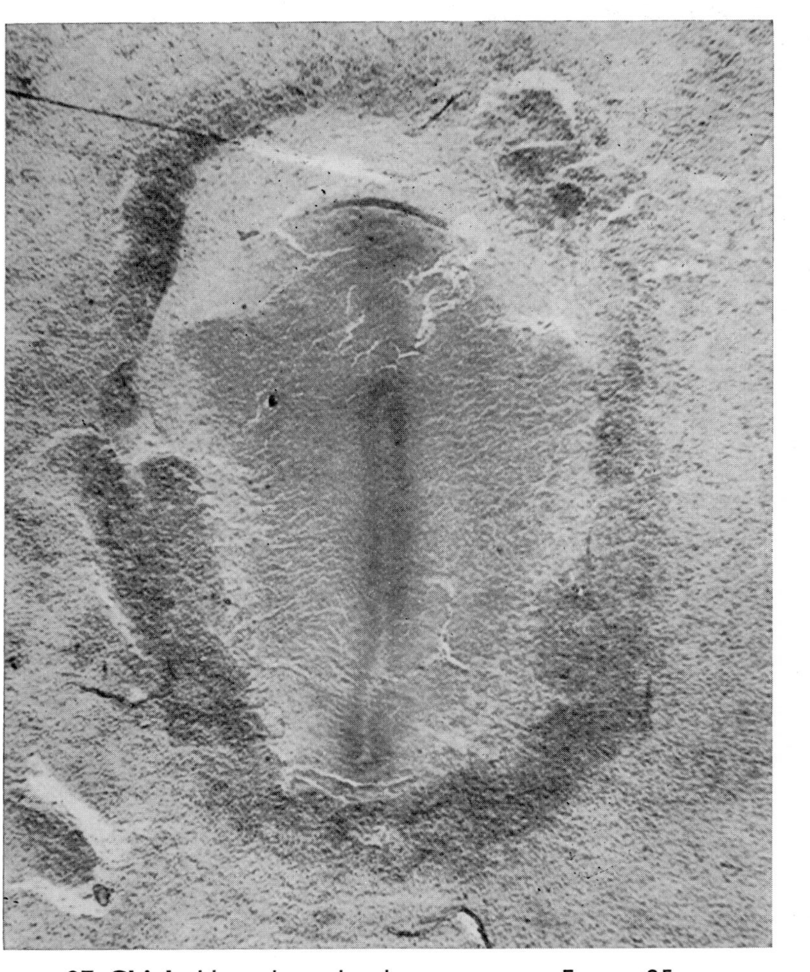

27. **Chick:** blastoderm, head-process stage, E. *mag*. 25×

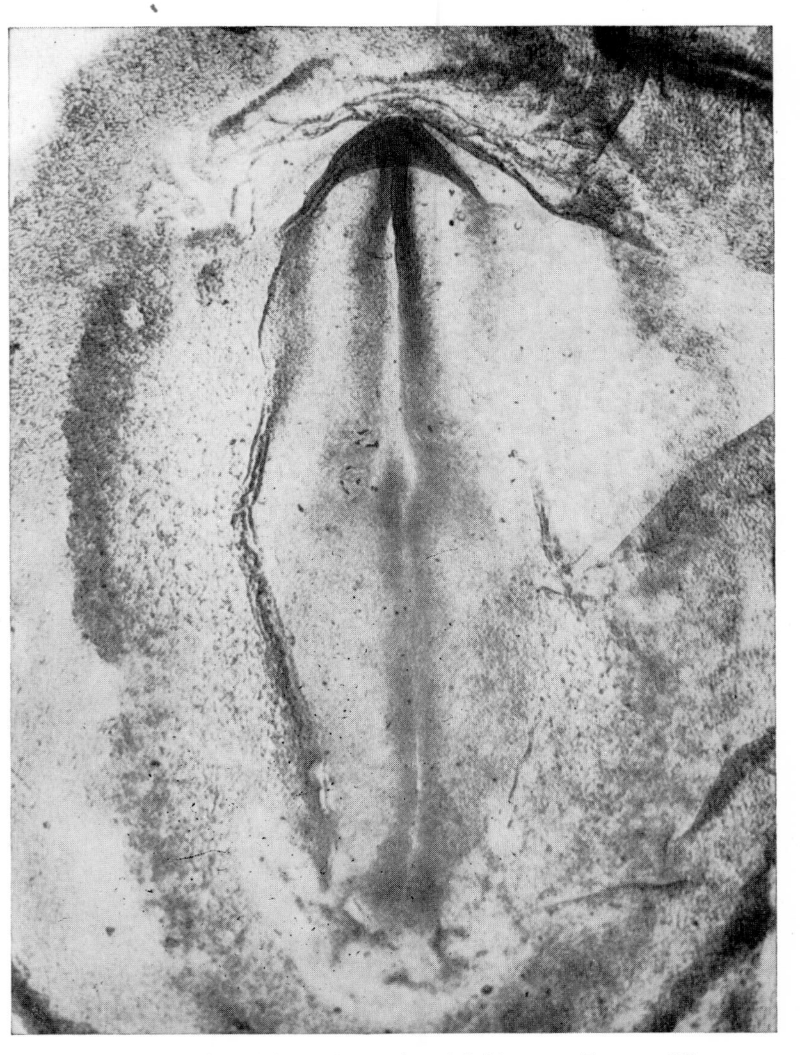

28. **Chick:** blastoderm, head-fold stage, E. *mag*. 25×

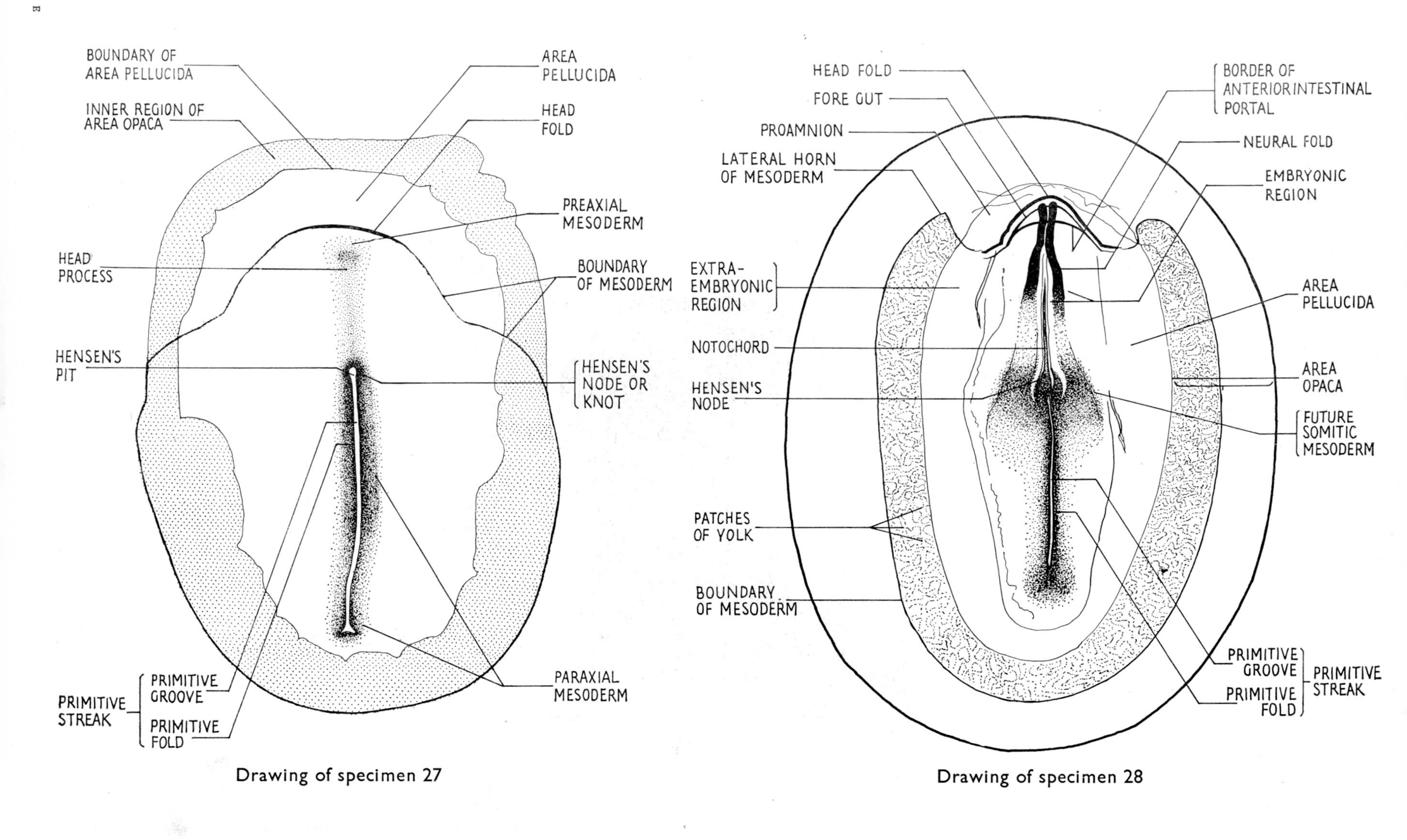
BOUNDARY OF AREA PELLUCIDA
INNER REGION OF AREA OPACA
HEAD PROCESS
HENSEN'S PIT
PRIMITIVE STREAK
PRIMITIVE GROOVE
PRIMITIVE FOLD
AREA PELLUCIDA
HEAD FOLD
PREAXIAL MESODERM
BOUNDARY OF MESODERM
HENSEN'S NODE OR KNOT
PARAXIAL MESODERM

Drawing of specimen 27

HEAD FOLD
FORE GUT
PROAMNION
LATERAL HORN OF MESODERM
EXTRA-EMBRYONIC REGION
NOTOCHORD
HENSEN'S NODE
PATCHES OF YOLK
BOUNDARY OF MESODERM
BORDER OF ANTERIOR INTESTINAL PORTAL
NEURAL FOLD
EMBRYONIC REGION
AREA PELLUCIDA
AREA OPACA
FUTURE SOMITIC MESODERM
PRIMITIVE GROOVE
PRIMITIVE FOLD
PRIMITIVE STREAK

Drawing of specimen 28

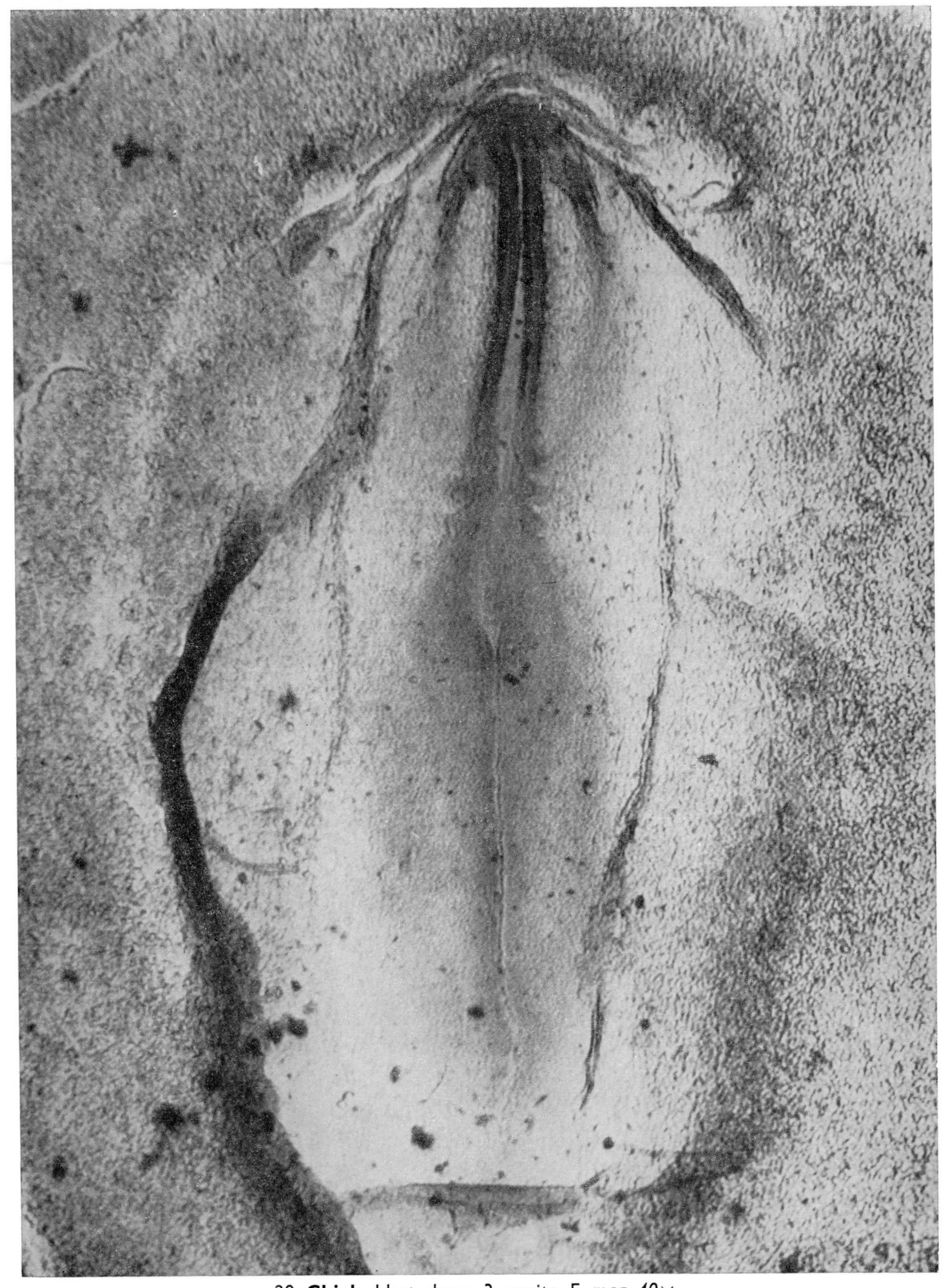

29. **Chick:** blastoderm, 3-somite, E. *mag.* 40×

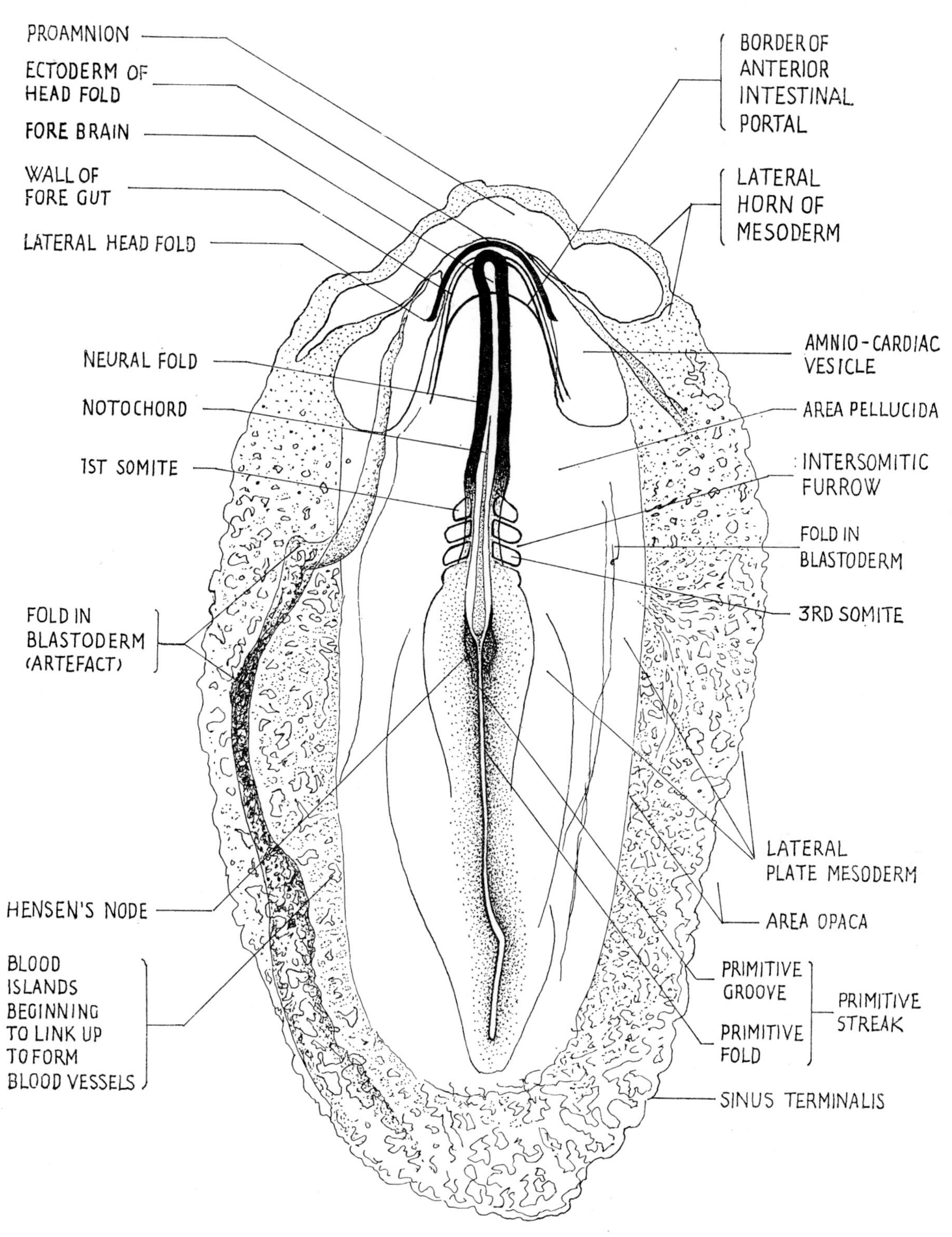
PROAMNION
ECTODERM OF HEAD FOLD
FORE BRAIN
WALL OF FORE GUT
LATERAL HEAD FOLD
NEURAL FOLD
NOTOCHORD
1ST SOMITE
FOLD IN BLASTODERM (ARTEFACT)
HENSEN'S NODE
BLOOD ISLANDS BEGINNING TO LINK UP TO FORM BLOOD VESSELS
BORDER OF ANTERIOR INTESTINAL PORTAL
LATERAL HORN OF MESODERM
AMNIO-CARDIAC VESICLE
AREA PELLUCIDA
INTERSOMITIC FURROW
FOLD IN BLASTODERM
3RD SOMITE
LATERAL PLATE MESODERM
AREA OPACA
PRIMITIVE GROOVE
PRIMITIVE FOLD
PRIMITIVE STREAK
SINUS TERMINALIS

Drawing of specimen 29

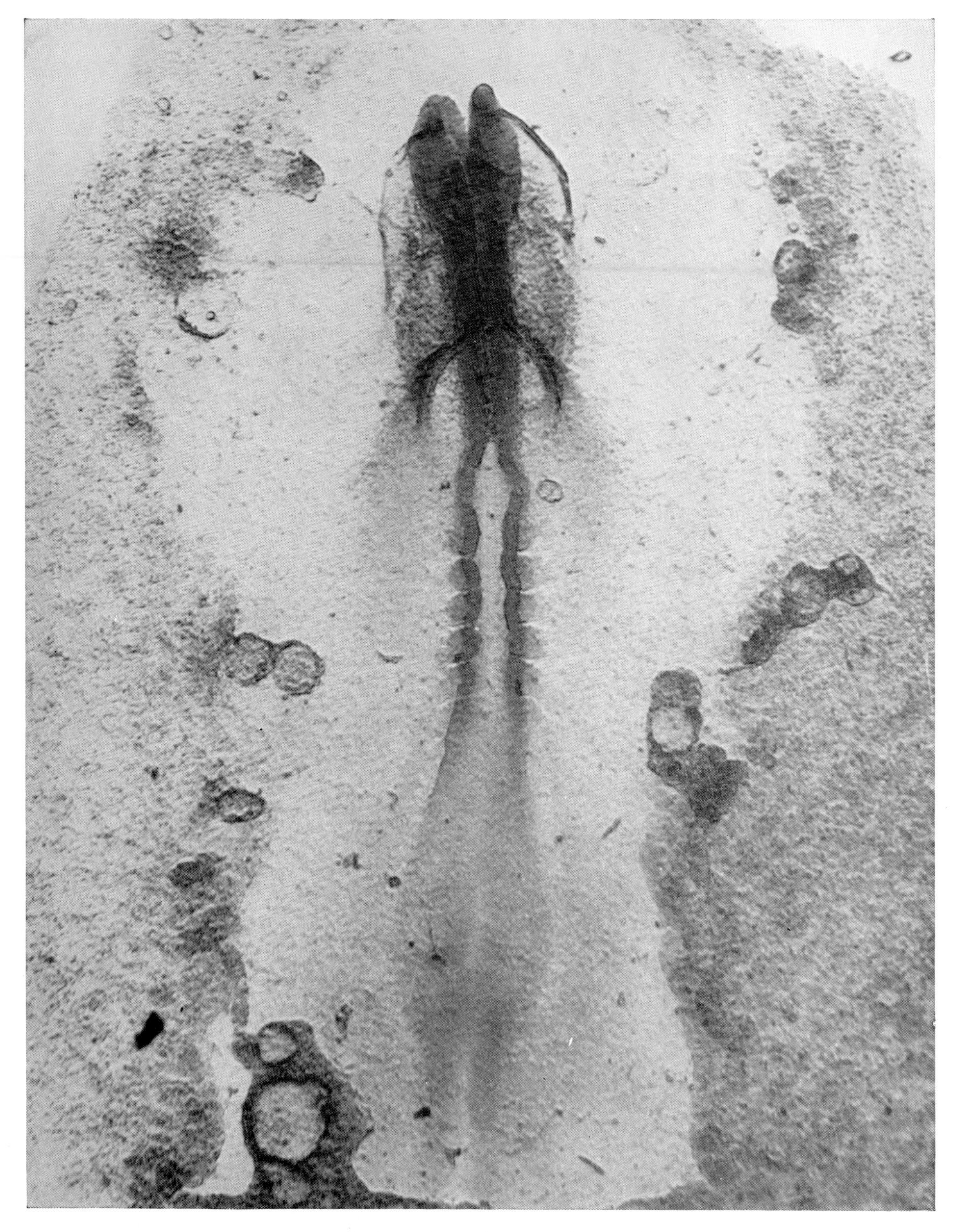

30. **Chick:** blastoderm, 6-somite, E. *mag. 40×*

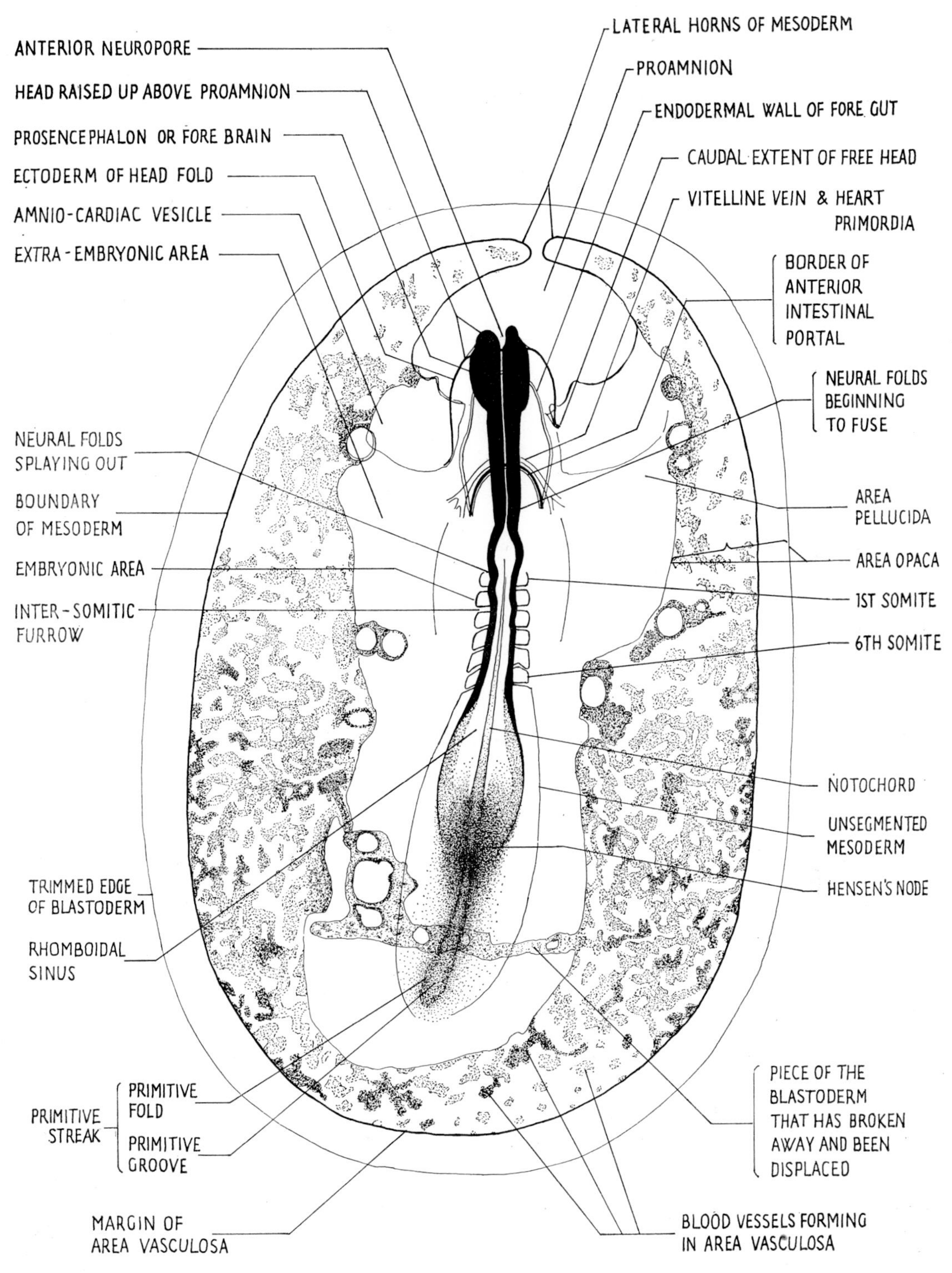
LATERAL HORNS OF MESODERM
ANTERIOR NEUROPORE
PROAMNION
HEAD RAISED UP ABOVE PROAMNION
ENDODERMAL WALL OF FORE GUT
PROSENCEPHALON OR FORE BRAIN
CAUDAL EXTENT OF FREE HEAD
ECTODERM OF HEAD FOLD
VITELLINE VEIN & HEART PRIMORDIA
AMNIO-CARDIAC VESICLE
EXTRA-EMBRYONIC AREA
BORDER OF ANTERIOR INTESTINAL PORTAL
NEURAL FOLDS BEGINNING TO FUSE
NEURAL FOLDS SPLAYING OUT
BOUNDARY OF MESODERM
AREA PELLUCIDA
AREA OPACA
EMBRYONIC AREA
1ST SOMITE
INTER-SOMITIC FURROW
6TH SOMITE
NOTOCHORD
UNSEGMENTED MESODERM
HENSEN'S NODE
TRIMMED EDGE OF BLASTODERM
RHOMBOIDAL SINUS
PRIMITIVE STREAK
PRIMITIVE FOLD
PRIMITIVE GROOVE
PIECE OF THE BLASTODERM THAT HAS BROKEN AWAY AND BEEN DISPLACED
MARGIN OF AREA VASCULOSA
BLOOD VESSELS FORMING IN AREA VASCULOSA

Drawing of specimen 30

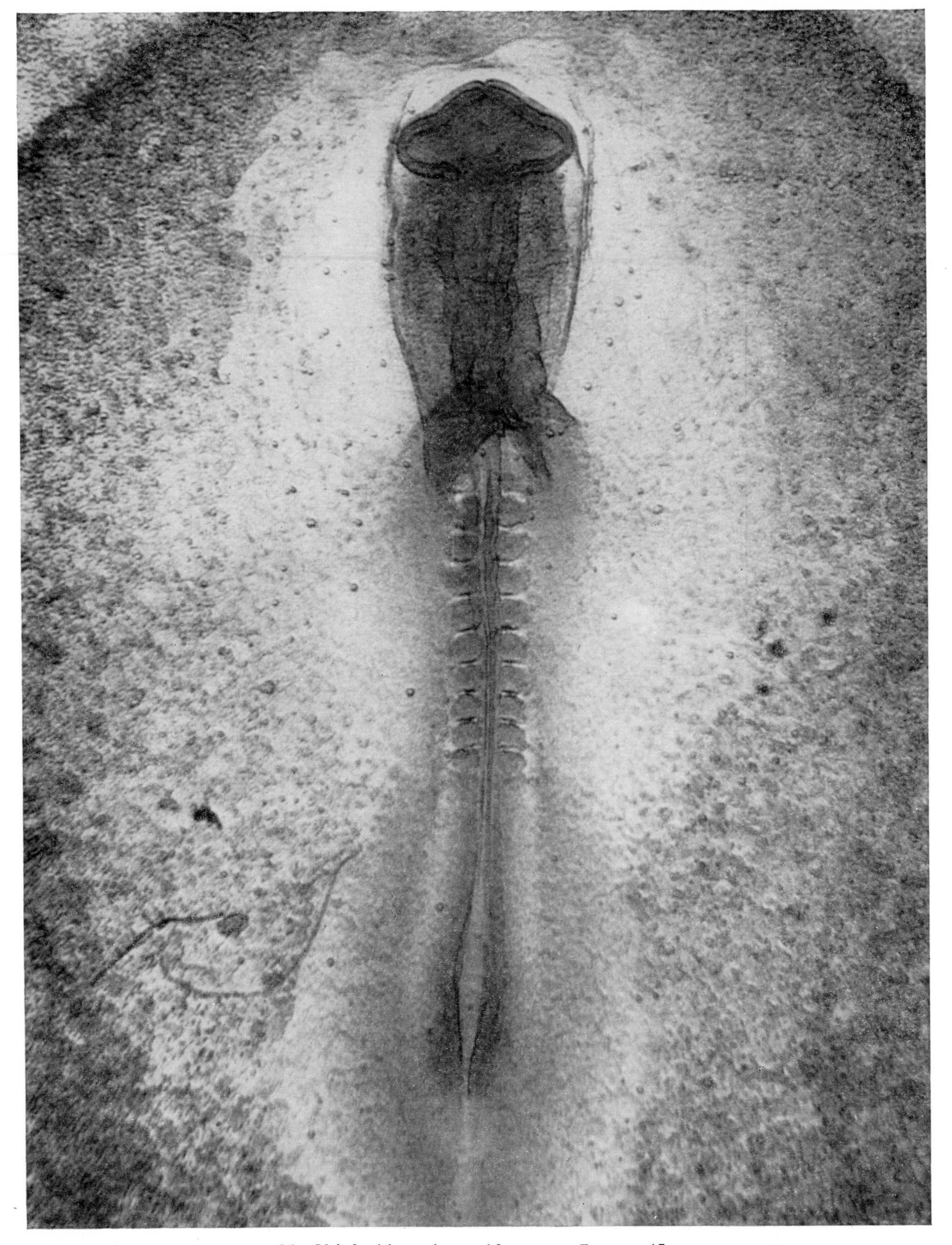

31. **Chick:** blastoderm, 10-somite, E. *mag.* 45×

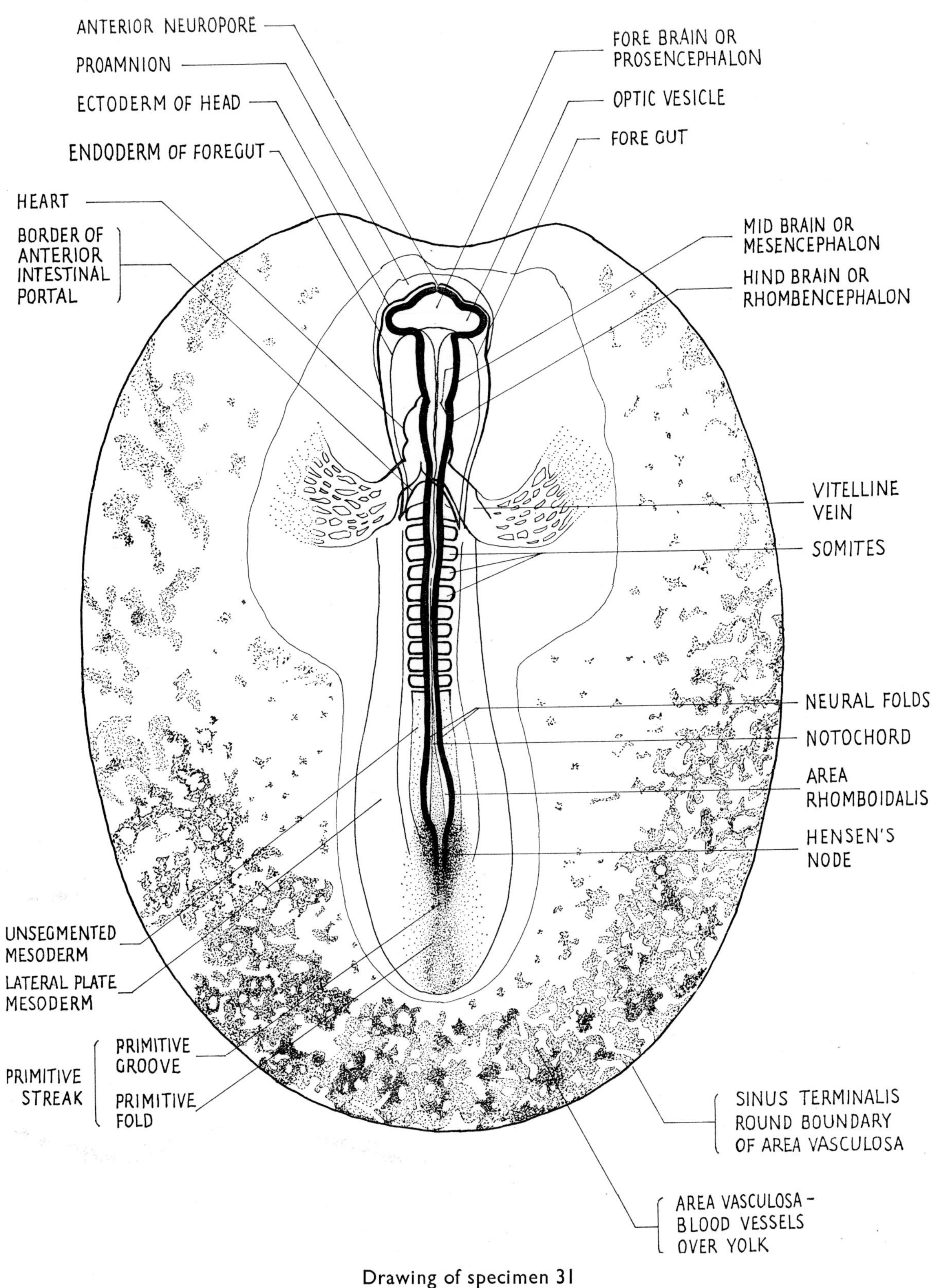
ANTERIOR NEUROPORE
PROAMNION
ECTODERM OF HEAD
ENDODERM OF FOREGUT
HEART
BORDER OF ANTERIOR INTESTINAL PORTAL
FORE BRAIN OR PROSENCEPHALON
OPTIC VESICLE
FORE GUT
MID BRAIN OR MESENCEPHALON
HIND BRAIN OR RHOMBENCEPHALON
VITELLINE VEIN
SOMITES
NEURAL FOLDS
NOTOCHORD
AREA RHOMBOIDALIS
HENSEN'S NODE
UNSEGMENTED MESODERM
LATERAL PLATE MESODERM
PRIMITIVE STREAK
PRIMITIVE GROOVE
PRIMITIVE FOLD
SINUS TERMINALIS ROUND BOUNDARY OF AREA VASCULOSA
AREA VASCULOSA - BLOOD VESSELS OVER YOLK

Drawing of specimen 31

(Drawn from ventral aspect; photograph is of dorsal aspect)

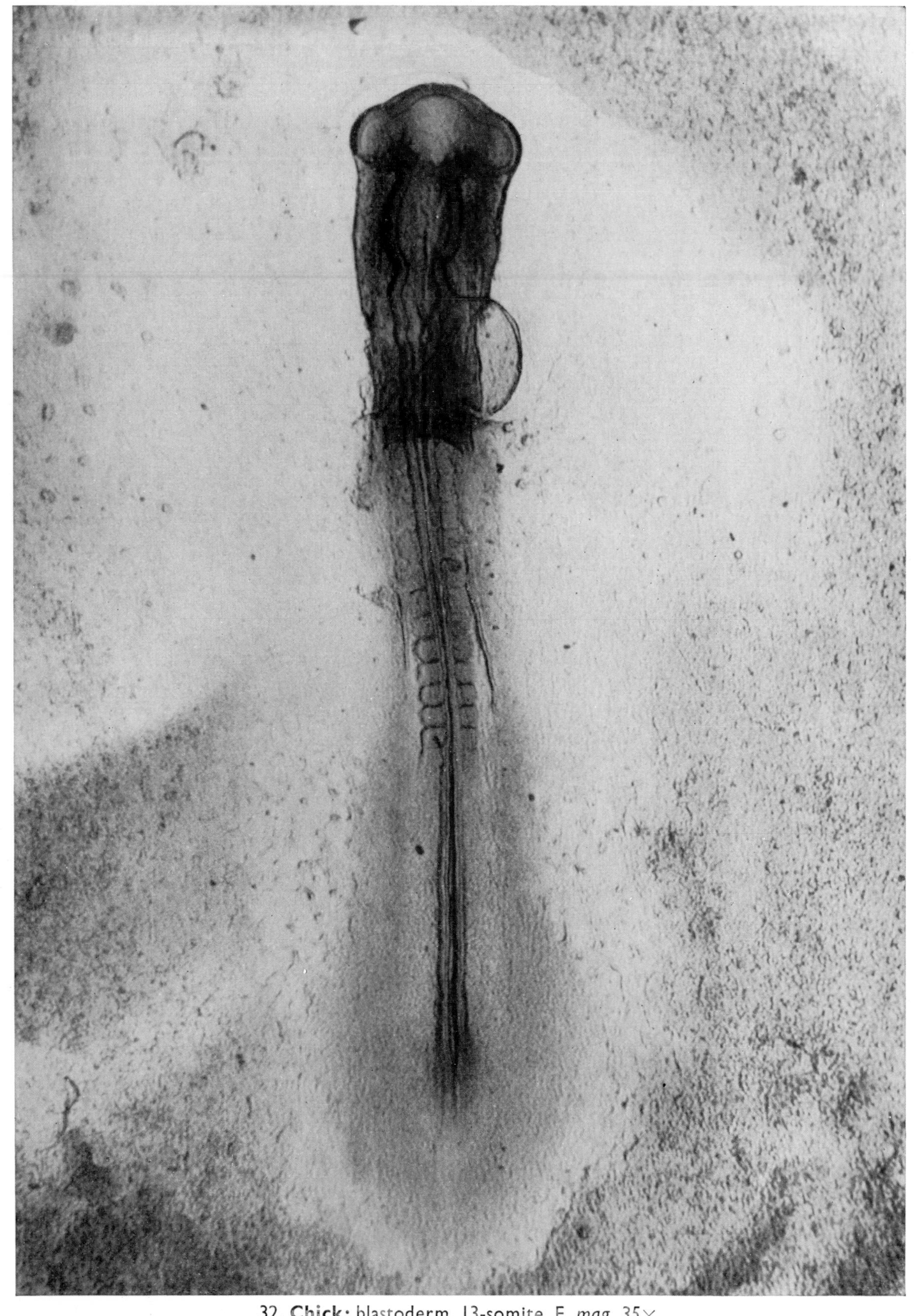

32. **Chick:** blastoderm, 13-somite, E. *mag.* 35×

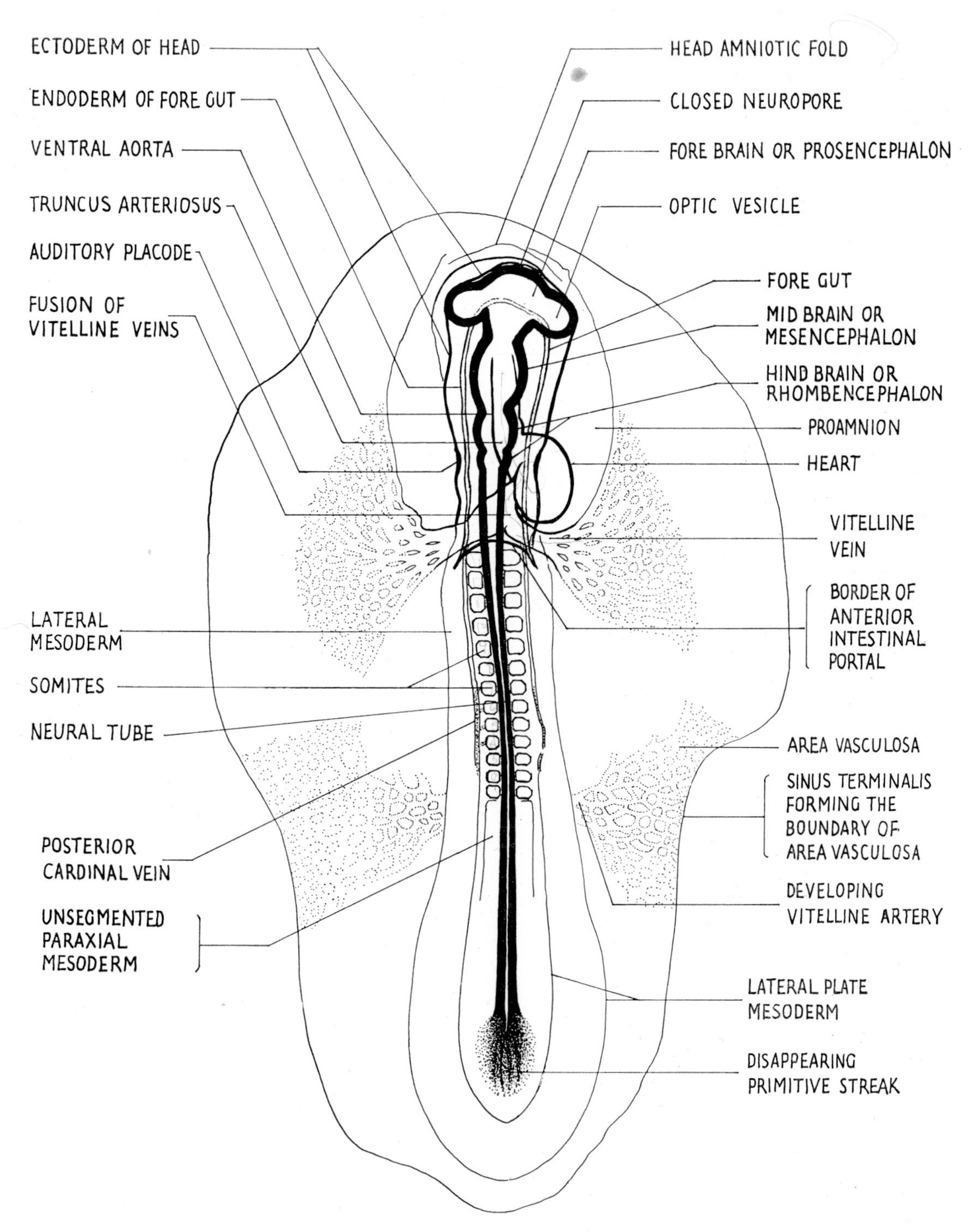
ECTODERM OF HEAD
ENDODERM OF FORE GUT
VENTRAL AORTA
TRUNCUS ARTERIOSUS
AUDITORY PLACODE
FUSION OF VITELLINE VEINS
LATERAL MESODERM
SOMITES
NEURAL TUBE
POSTERIOR CARDINAL VEIN
UNSEGMENTED PARAXIAL MESODERM
HEAD AMNIOTIC FOLD
CLOSED NEUROPORE
FORE BRAIN OR PROSENCEPHALON
OPTIC VESICLE
FORE GUT
MID BRAIN OR MESENCEPHALON
HIND BRAIN OR RHOMBENCEPHALON
PROAMNION
HEART
VITELLINE VEIN
BORDER OF ANTERIOR INTESTINAL PORTAL
AREA VASCULOSA
SINUS TERMINALIS FORMING THE BOUNDARY OF AREA VASCULOSA
DEVELOPING VITELLINE ARTERY
LATERAL PLATE MESODERM
DISAPPEARING PRIMITIVE STREAK

Drawing of specimen 32

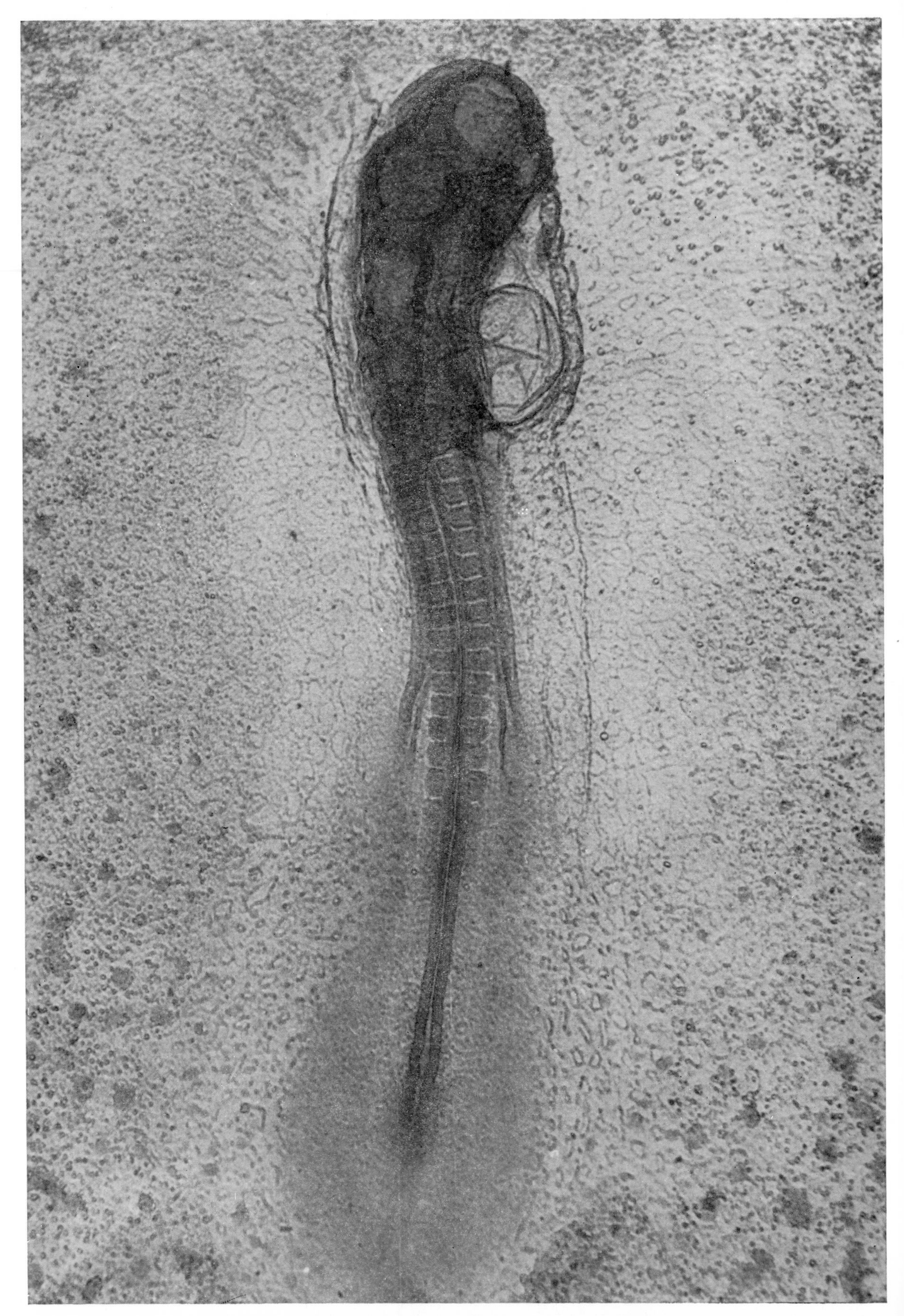

33. **Chick**: blastoderm, 17-somite, E. *mag.* 30×

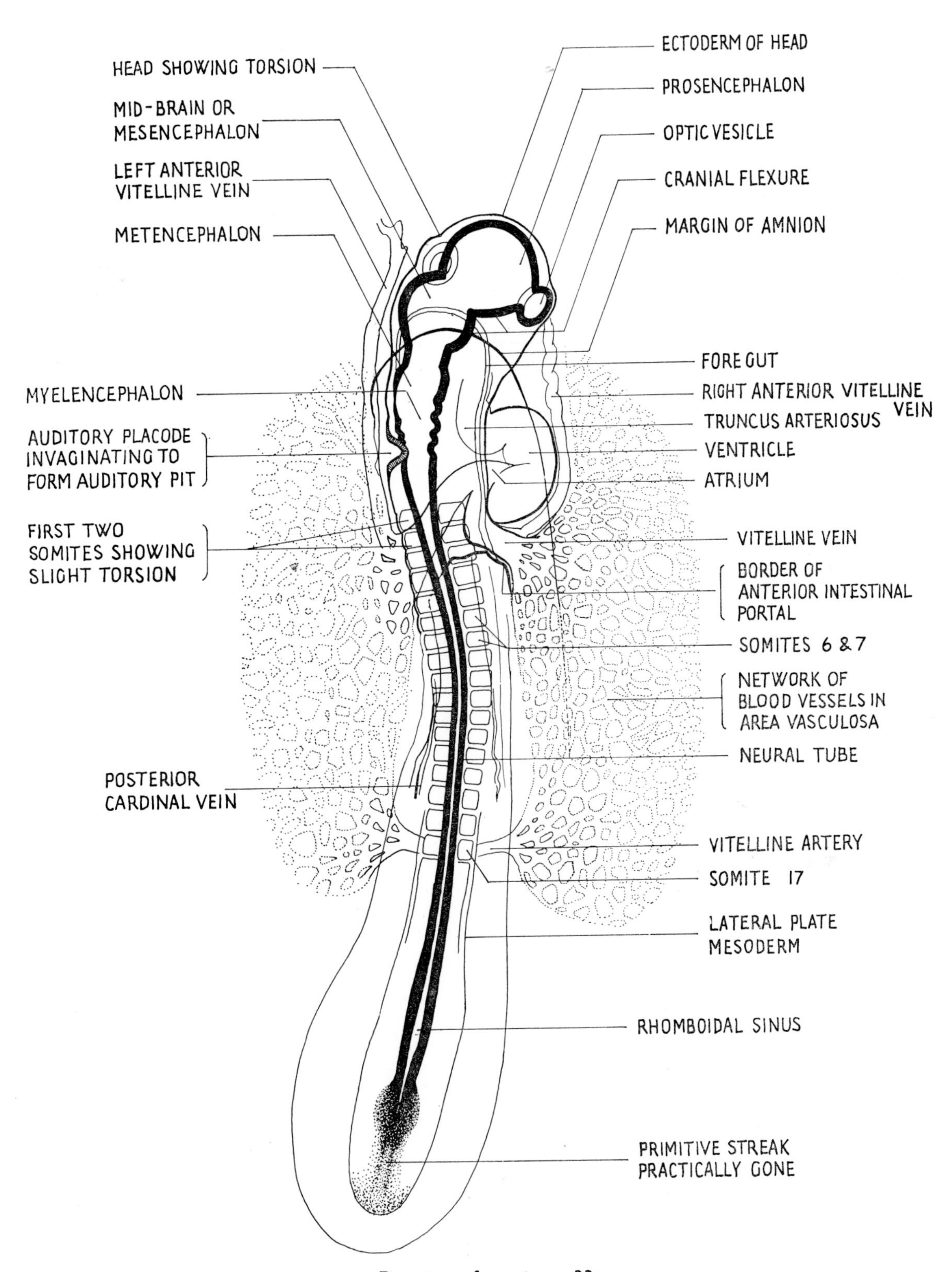
HEAD SHOWING TORSION
MID-BRAIN OR MESENCEPHALON
LEFT ANTERIOR VITELLINE VEIN
METENCEPHALON
MYELENCEPHALON
AUDITORY PLACODE INVAGINATING TO FORM AUDITORY PIT
FIRST TWO SOMITES SHOWING SLIGHT TORSION
POSTERIOR CARDINAL VEIN
ECTODERM OF HEAD
PROSENCEPHALON
OPTIC VESICLE
CRANIAL FLEXURE
MARGIN OF AMNION
FORE GUT
RIGHT ANTERIOR VITELLINE VEIN
TRUNCUS ARTERIOSUS
VENTRICLE
ATRIUM
VITELLINE VEIN
BORDER OF ANTERIOR INTESTINAL PORTAL
SOMITES 6 & 7
NETWORK OF BLOOD VESSELS IN AREA VASCULOSA
NEURAL TUBE
VITELLINE ARTERY
SOMITE 17
LATERAL PLATE MESODERM
RHOMBOIDAL SINUS
PRIMITIVE STREAK PRACTICALLY GONE

Drawing of specimen 33

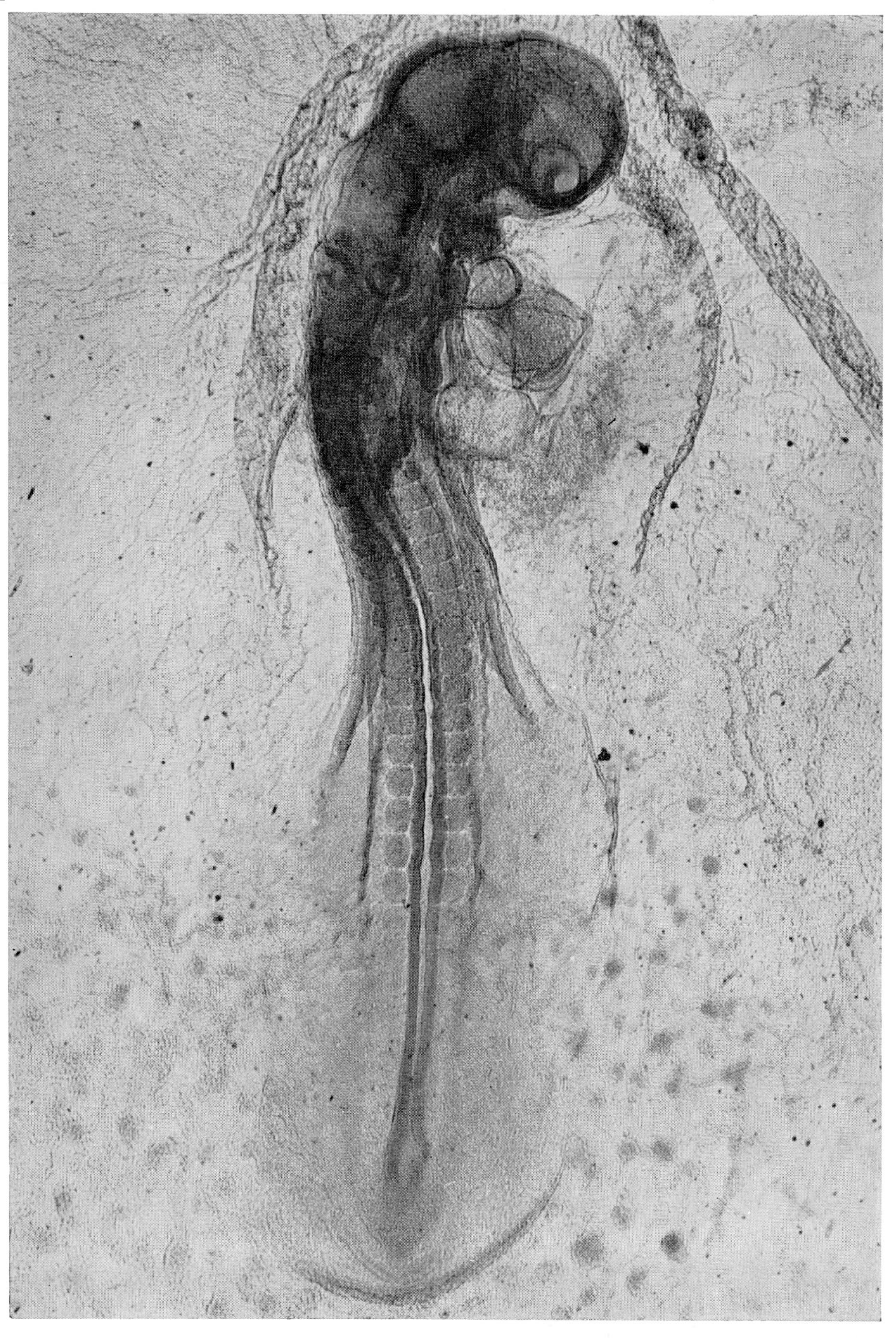

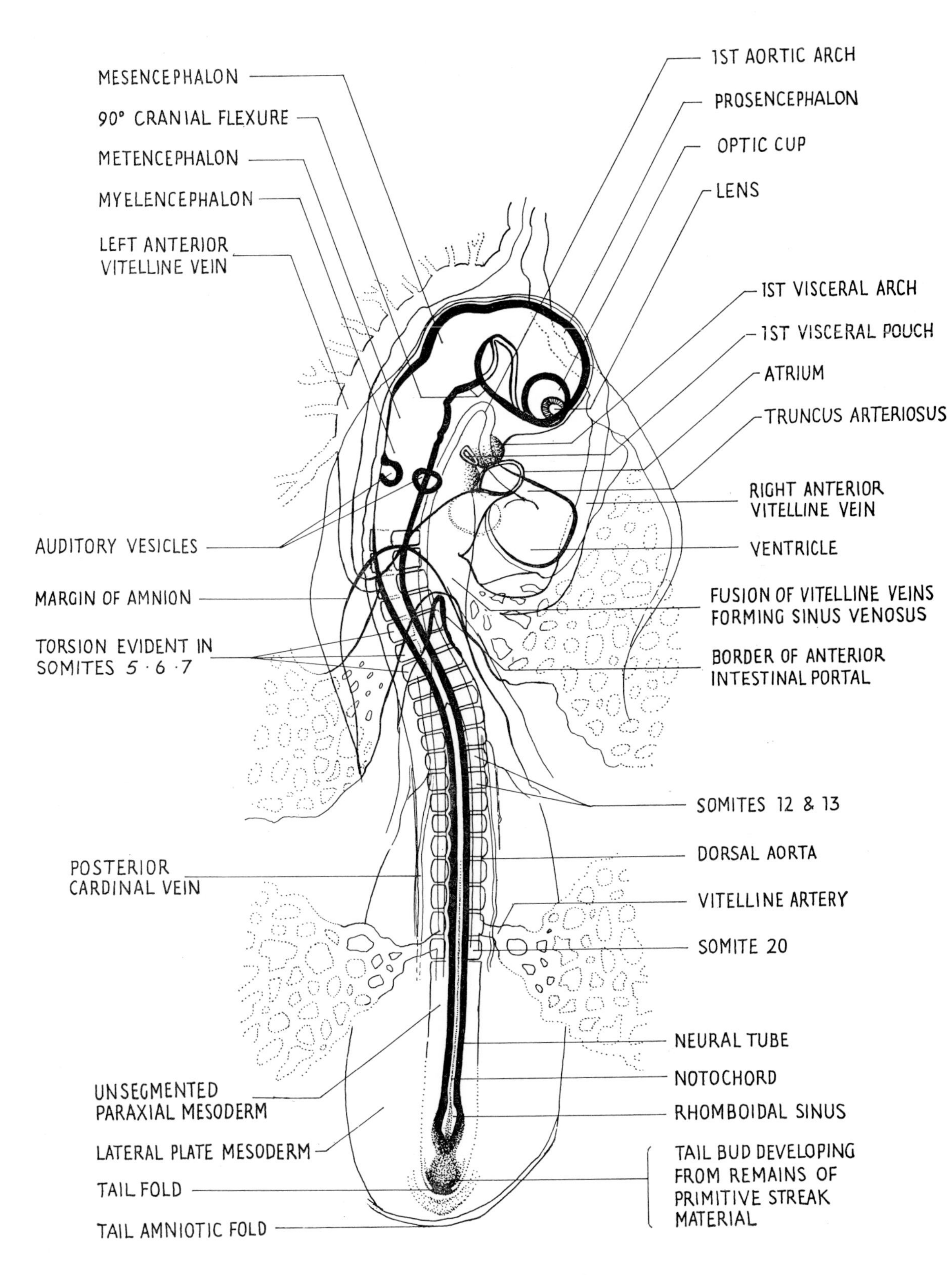
MESENCEPHALON
90° CRANIAL FLEXURE
METENCEPHALON
MYELENCEPHALON
LEFT ANTERIOR VITELLINE VEIN
AUDITORY VESICLES
MARGIN OF AMNION
TORSION EVIDENT IN SOMITES 5 · 6 · 7
POSTERIOR CARDINAL VEIN
UNSEGMENTED PARAXIAL MESODERM
LATERAL PLATE MESODERM
TAIL FOLD
TAIL AMNIOTIC FOLD
1ST AORTIC ARCH
PROSENCEPHALON
OPTIC CUP
LENS
1ST VISCERAL ARCH
1ST VISCERAL POUCH
ATRIUM
TRUNCUS ARTERIOSUS
RIGHT ANTERIOR VITELLINE VEIN
VENTRICLE
FUSION OF VITELLINE VEINS FORMING SINUS VENOSUS
BORDER OF ANTERIOR INTESTINAL PORTAL
SOMITES 12 & 13
DORSAL AORTA
VITELLINE ARTERY
SOMITE 20
NEURAL TUBE
NOTOCHORD
RHOMBOIDAL SINUS
TAIL BUD DEVELOPING FROM REMAINS OF PRIMITIVE STREAK MATERIAL

Drawing of specimen 34

(*Left*) 34. **Chick:** blastoderm, 20-somite, E. *mag. 40*×

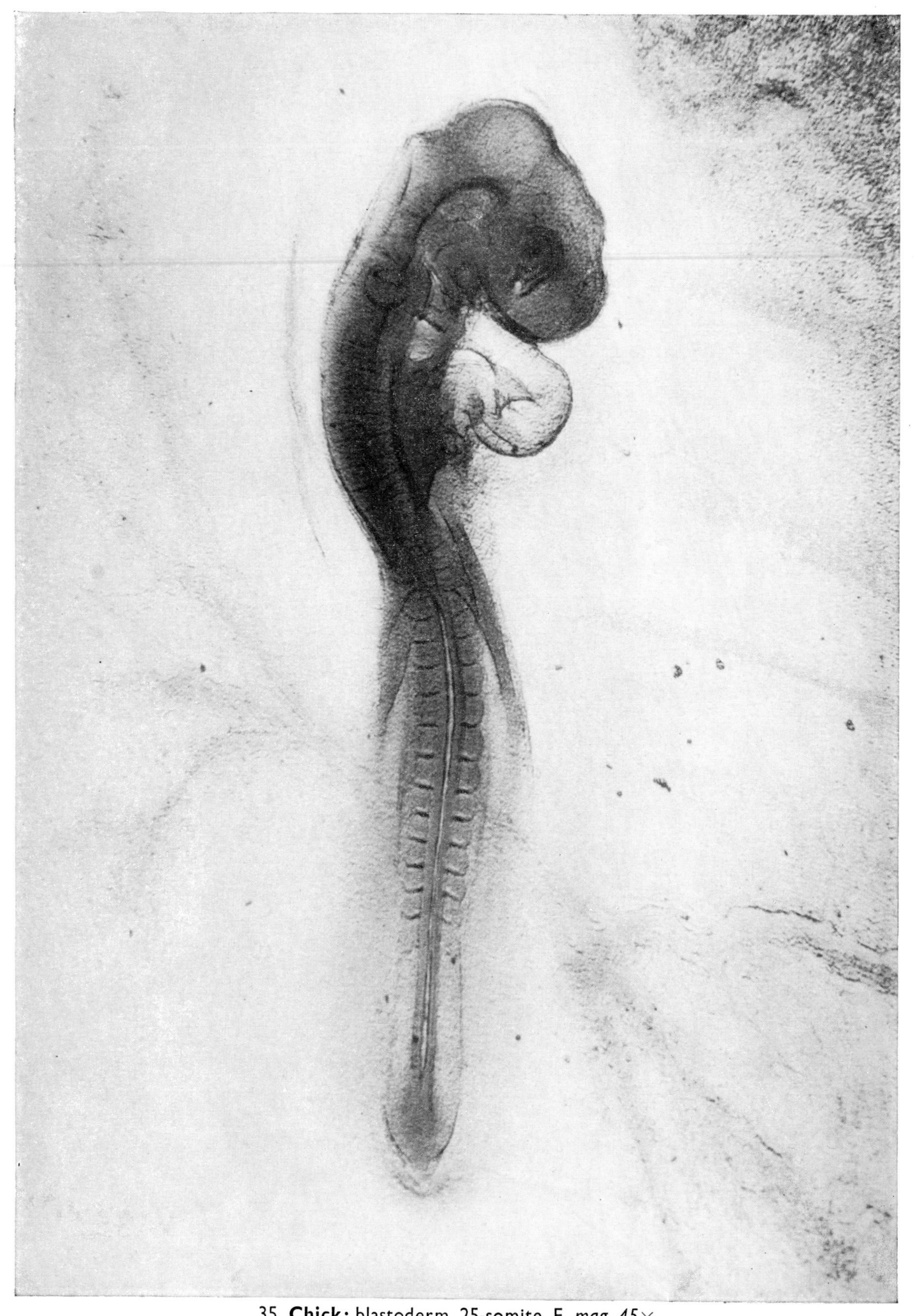

35. **Chick:** blastoderm, 25-somite, E. *mag.* 45×

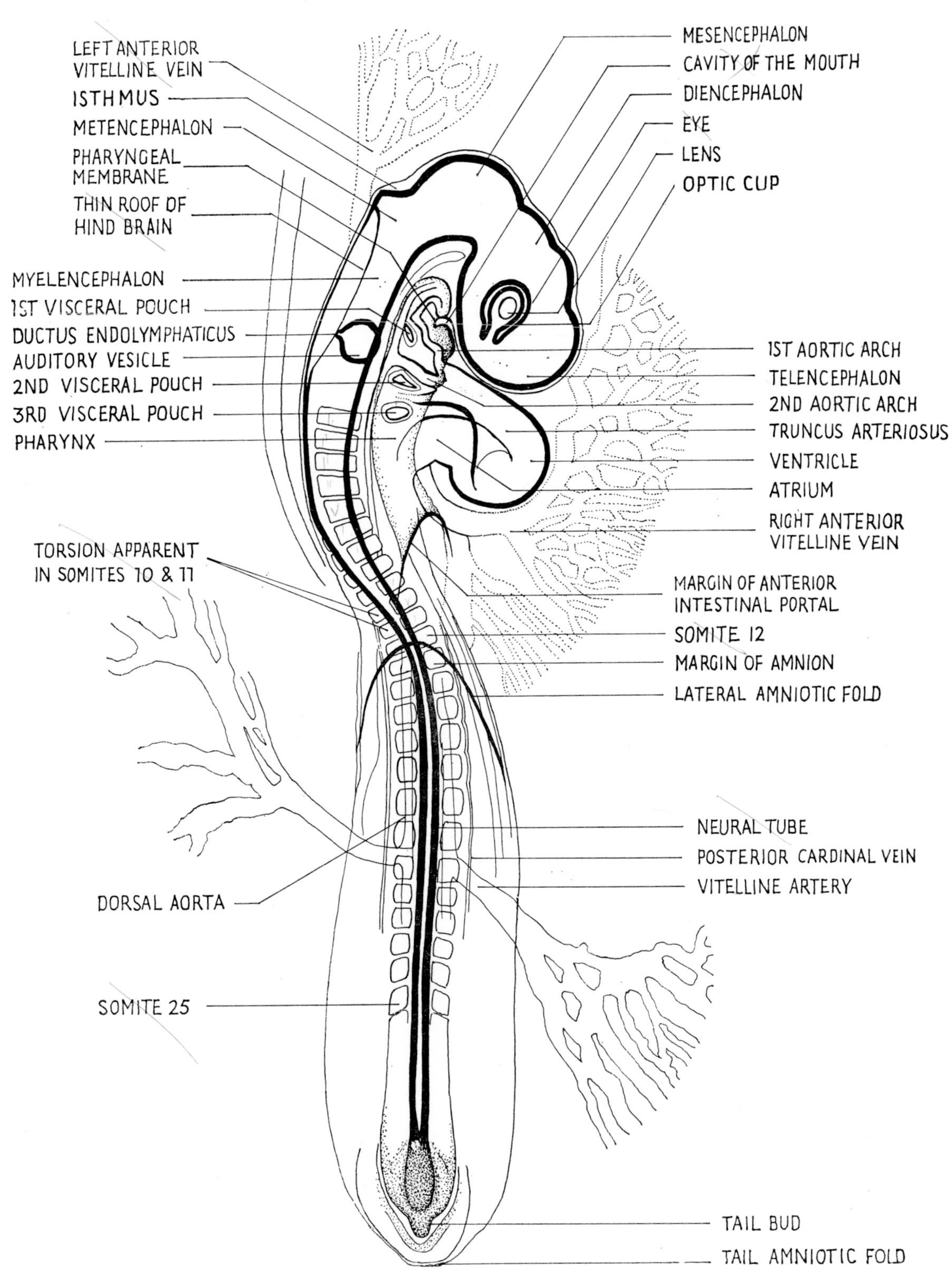
LEFT ANTERIOR VITELLINE VEIN
ISTHMUS
METENCEPHALON
PHARYNGEAL MEMBRANE
THIN ROOF OF HIND BRAIN
MYELENCEPHALON
1ST VISCERAL POUCH
DUCTUS ENDOLYMPHATICUS
AUDITORY VESICLE
2ND VISCERAL POUCH
3RD VISCERAL POUCH
PHARYNX
TORSION APPARENT IN SOMITES 10 & 11
DORSAL AORTA
SOMITE 25
MESENCEPHALON
CAVITY OF THE MOUTH
DIENCEPHALON
EYE
LENS
OPTIC CUP
1ST AORTIC ARCH
TELENCEPHALON
2ND AORTIC ARCH
TRUNCUS ARTERIOSUS
VENTRICLE
ATRIUM
RIGHT ANTERIOR VITELLINE VEIN
MARGIN OF ANTERIOR INTESTINAL PORTAL
SOMITE 12
MARGIN OF AMNION
LATERAL AMNIOTIC FOLD
NEURAL TUBE
POSTERIOR CARDINAL VEIN
VITELLINE ARTERY
TAIL BUD
TAIL AMNIOTIC FOLD

Drawing of specimen 35

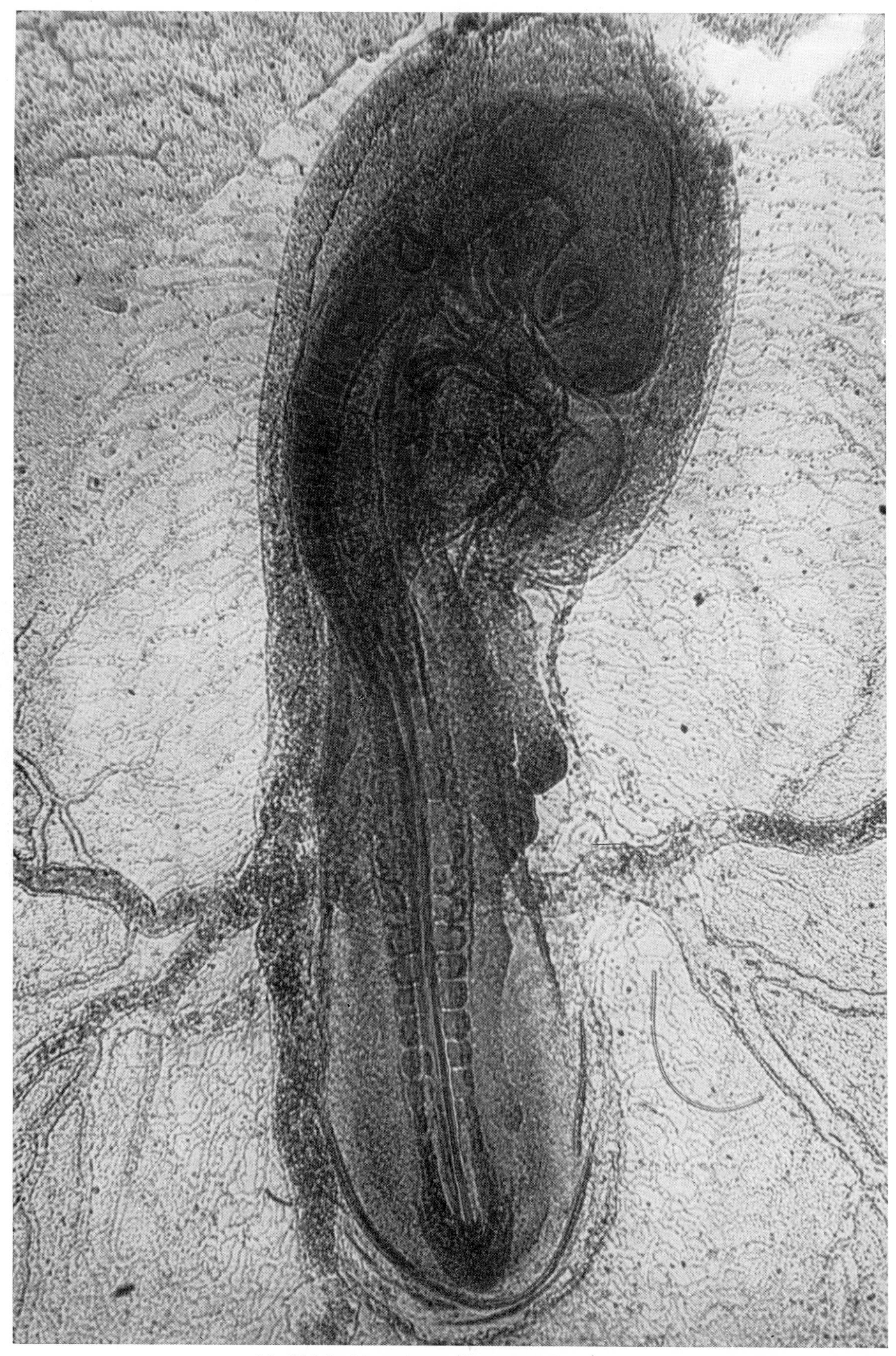

36. **Chick:** blastoderm, 30-somite, E. *mdg*. 25×

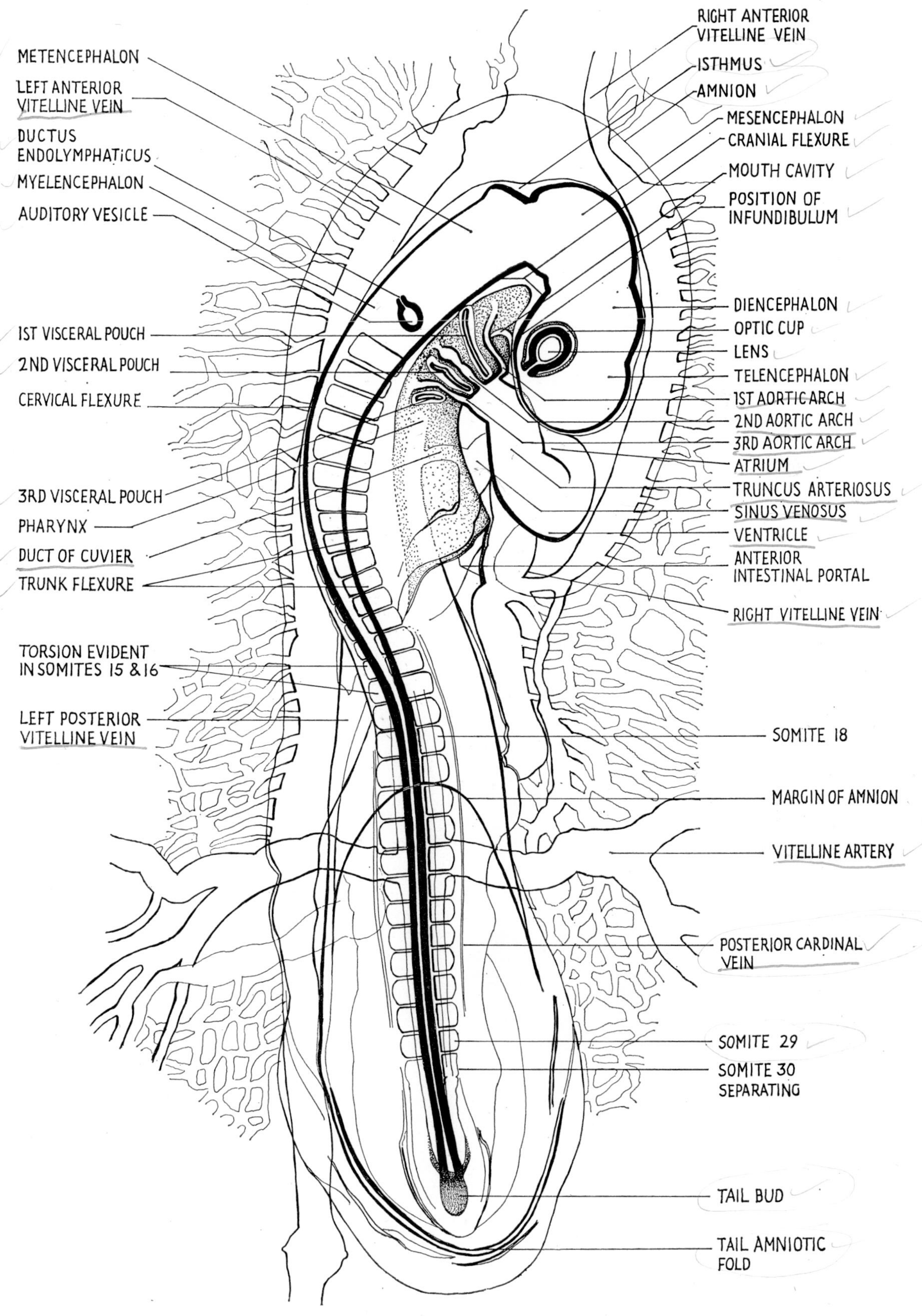
METENCEPHALON
LEFT ANTERIOR VITELLINE VEIN
DUCTUS ENDOLYMPHATICUS
MYELENCEPHALON
AUDITORY VESICLE
1ST VISCERAL POUCH
2ND VISCERAL POUCH
CERVICAL FLEXURE
3RD VISCERAL POUCH
PHARYNX
DUCT OF CUVIER
TRUNK FLEXURE
TORSION EVIDENT IN SOMITES 15 &16
LEFT POSTERIOR VITELLINE VEIN
RIGHT ANTERIOR VITELLINE VEIN
ISTHMUS
AMNION
MESENCEPHALON
CRANIAL FLEXURE
MOUTH CAVITY
POSITION OF INFUNDIBULUM
DIENCEPHALON
OPTIC CUP
LENS
TELENCEPHALON
1ST AORTIC ARCH
2ND AORTIC ARCH
3RD AORTIC ARCH
ATRIUM
TRUNCUS ARTERIOSUS
SINUS VENOSUS
VENTRICLE
ANTERIOR INTESTINAL PORTAL
RIGHT VITELLINE VEIN
SOMITE 18
MARGIN OF AMNION
VITELLINE ARTERY
POSTERIOR CARDINAL VEIN
SOMITE 29
SOMITE 30 SEPARATING
TAIL BUD
TAIL AMNIOTIC FOLD

Drawing of specimen 36

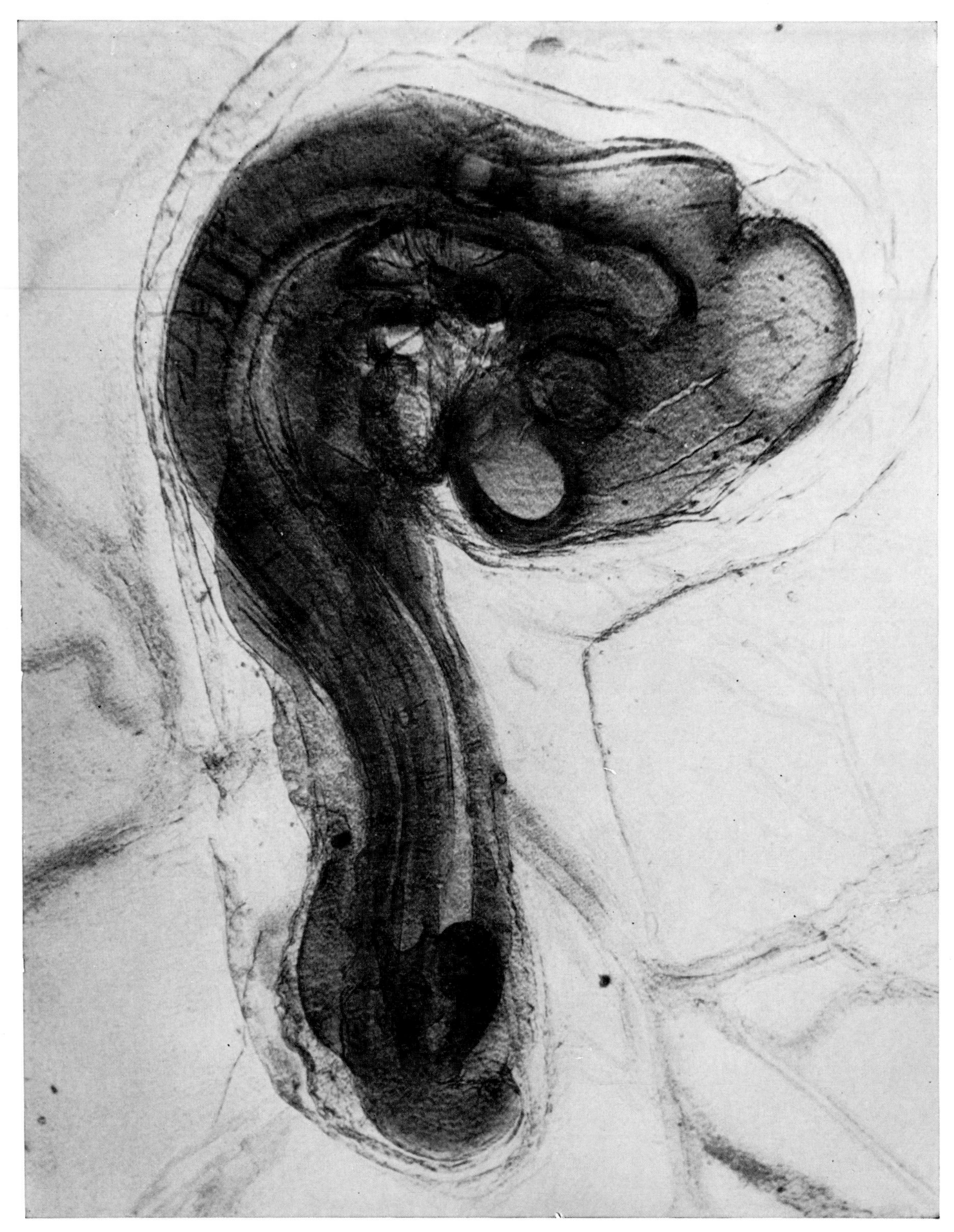

37. **Chick:** blastoderm, 35-somite, E. *mag.* 30×

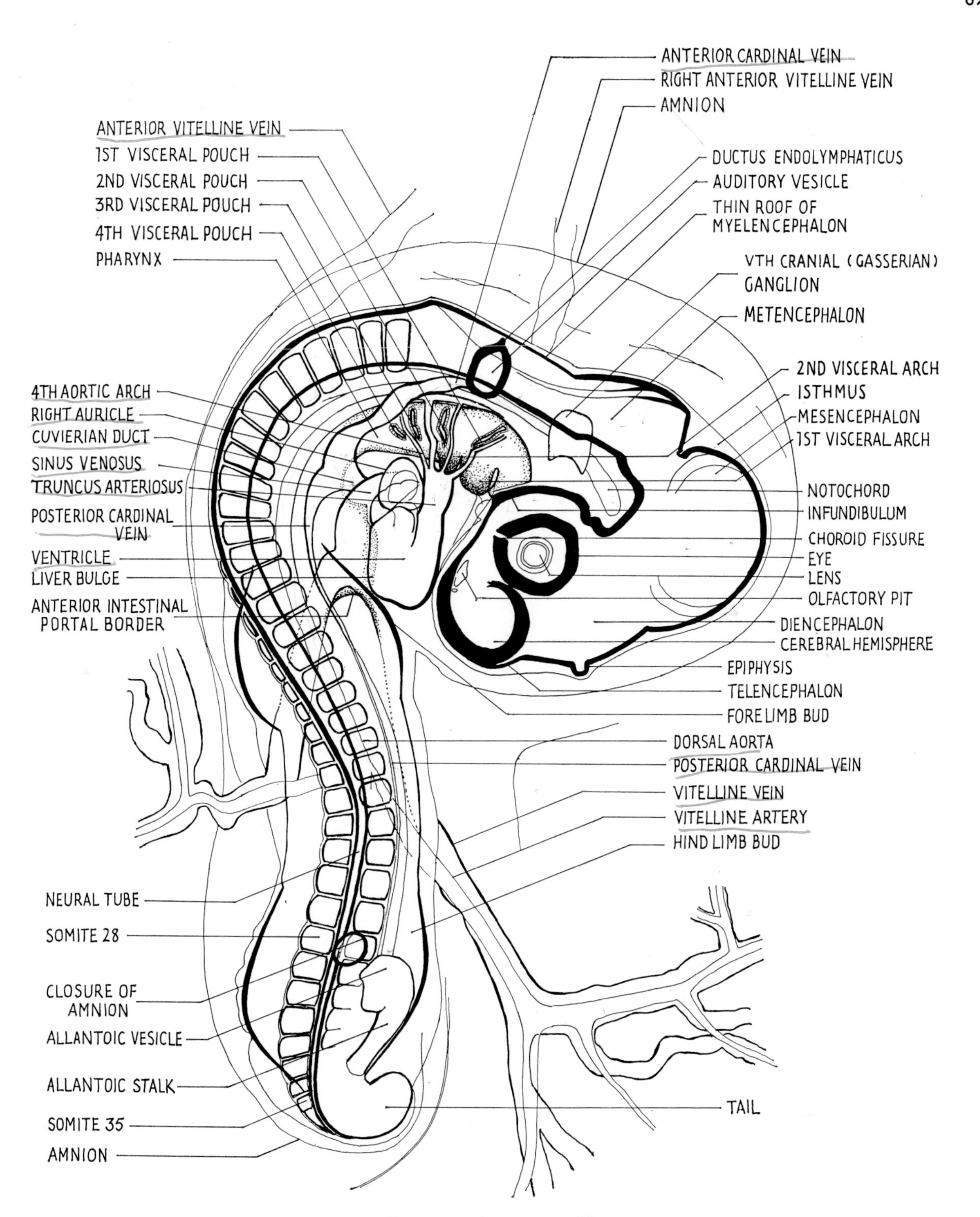

ANTERIOR CARDINAL VEIN
RIGHT ANTERIOR VITELLINE VEIN
AMNION
ANTERIOR VITELLINE VEIN
1ST VISCERAL POUCH
2ND VISCERAL POUCH
3RD VISCERAL POUCH
4TH VISCERAL POUCH
PHARYNX
DUCTUS ENDOLYMPHATICUS
AUDITORY VESICLE
THIN ROOF OF MYELENCEPHALON
VTH CRANIAL (GASSERIAN) GANGLION
METENCEPHALON
2ND VISCERAL ARCH
ISTHMUS
MESENCEPHALON
1ST VISCERAL ARCH
4TH AORTIC ARCH
RIGHT AURICLE
CUVIERIAN DUCT
SINUS VENOSUS
TRUNCUS ARTERIOSUS
POSTERIOR CARDINAL VEIN
VENTRICLE
LIVER BULGE
ANTERIOR INTESTINAL PORTAL BORDER
NOTOCHORD
INFUNDIBULUM
CHOROID FISSURE
EYE
LENS
OLFACTORY PIT
DIENCEPHALON
CEREBRAL HEMISPHERE
EPIPHYSIS
TELENCEPHALON
FORE LIMB BUD
DORSAL AORTA
POSTERIOR CARDINAL VEIN
VITELLINE VEIN
VITELLINE ARTERY
HIND LIMB BUD
NEURAL TUBE
SOMITE 28
CLOSURE OF AMNION
ALLANTOIC VESICLE
ALLANTOIC STALK
SOMITE 35
AMNION
TAIL

Drawing of specimen 37

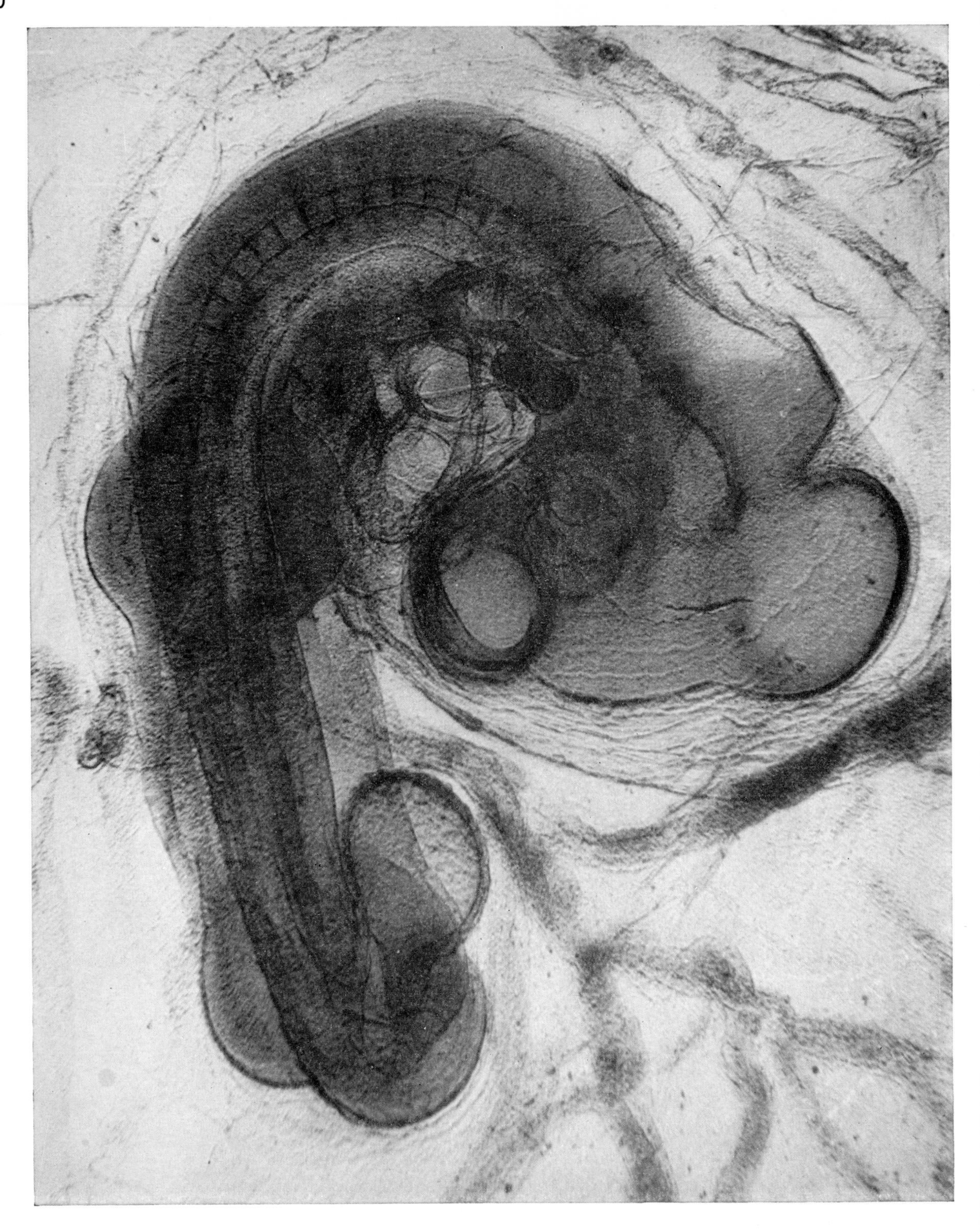

38. **Chick:** blastoderm, 40-somite, E. *mag. 30*×

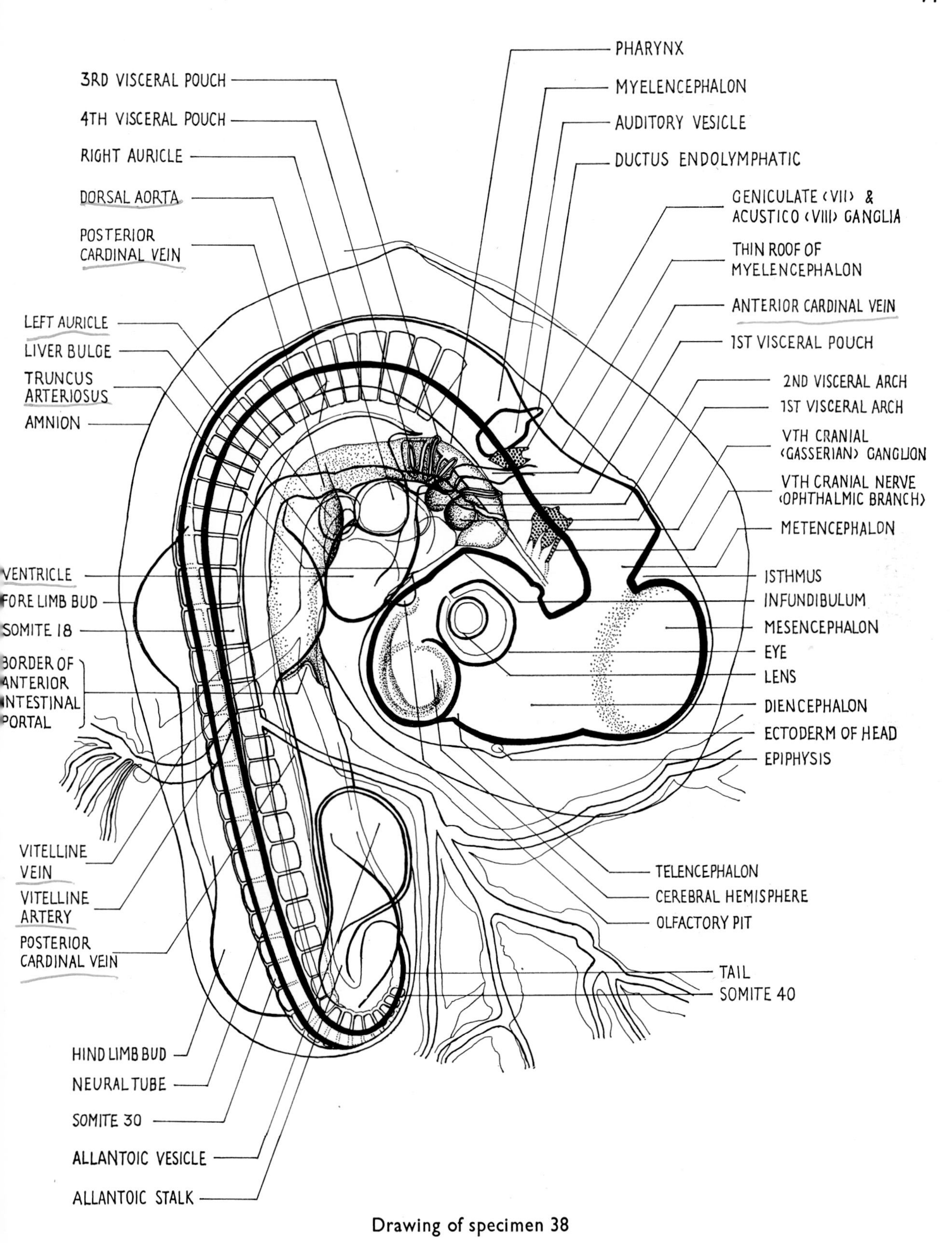

PHARYNX
MYELENCEPHALON
AUDITORY VESICLE
DUCTUS ENDOLYMPHATIC
GENICULATE (VII) & ACUSTICO (VIII) GANGLIA
THIN ROOF OF MYELENCEPHALON
ANTERIOR CARDINAL VEIN
1ST VISCERAL POUCH
2ND VISCERAL ARCH
1ST VISCERAL ARCH
VTH CRANIAL (GASSERIAN) GANGLION
VTH CRANIAL NERVE (OPHTHALMIC BRANCH)
METENCEPHALON
ISTHMUS
INFUNDIBULUM
MESENCEPHALON
EYE
LENS
DIENCEPHALON
ECTODERM OF HEAD
EPIPHYSIS
TELENCEPHALON
CEREBRAL HEMISPHERE
OLFACTORY PIT
TAIL
SOMITE 40
3RD VISCERAL POUCH
4TH VISCERAL POUCH
RIGHT AURICLE
DORSAL AORTA
POSTERIOR CARDINAL VEIN
LEFT AURICLE
LIVER BULGE
TRUNCUS ARTERIOSUS
AMNION
VENTRICLE
FORE LIMB BUD
SOMITE 18
BORDER OF ANTERIOR INTESTINAL PORTAL
VITELLINE VEIN
VITELLINE ARTERY
POSTERIOR CARDINAL VEIN
HIND LIMB BUD
NEURAL TUBE
SOMITE 30
ALLANTOIC VESICLE
ALLANTOIC STALK

Drawing of specimen 38

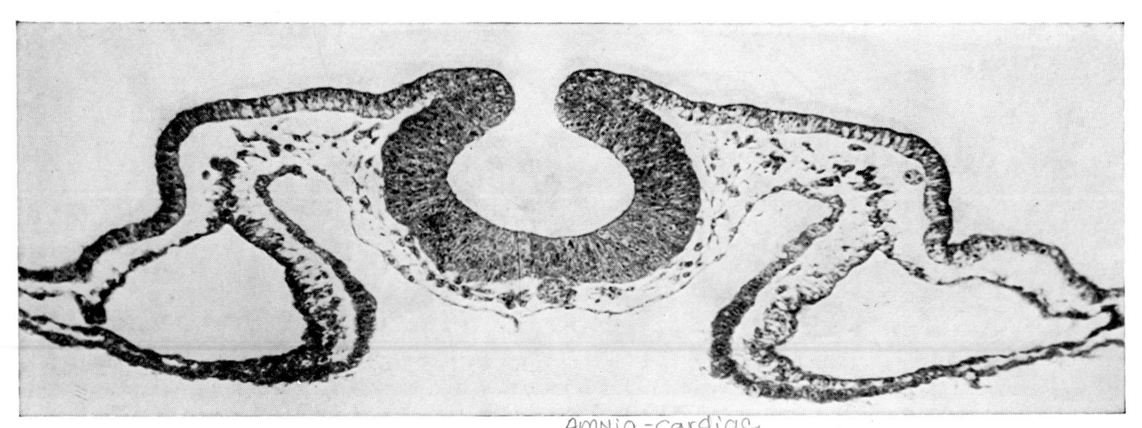

39. **Chick:** 6-somite stage, head region, T.S. *mag. 140×*

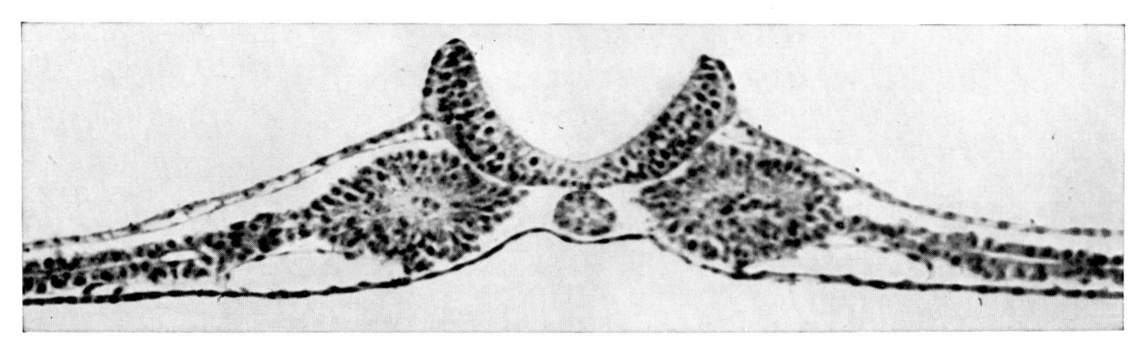

40. **Chick:** 6-somite stage, somitic region, T.S. *mag. 200×*

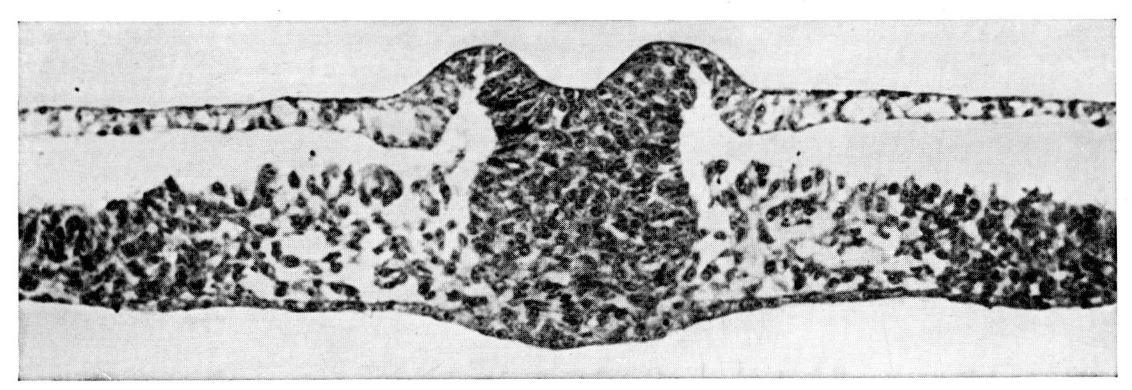

41. **Chick:** 6-somite stage, notochord, T.S. *mag. 225×*

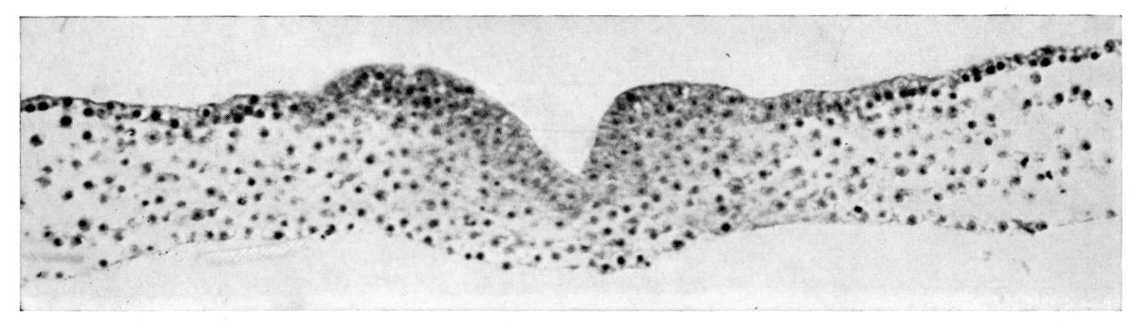

42. **Chick:** 6-somite stage, primitive streak, T.S. *mag. 200×*

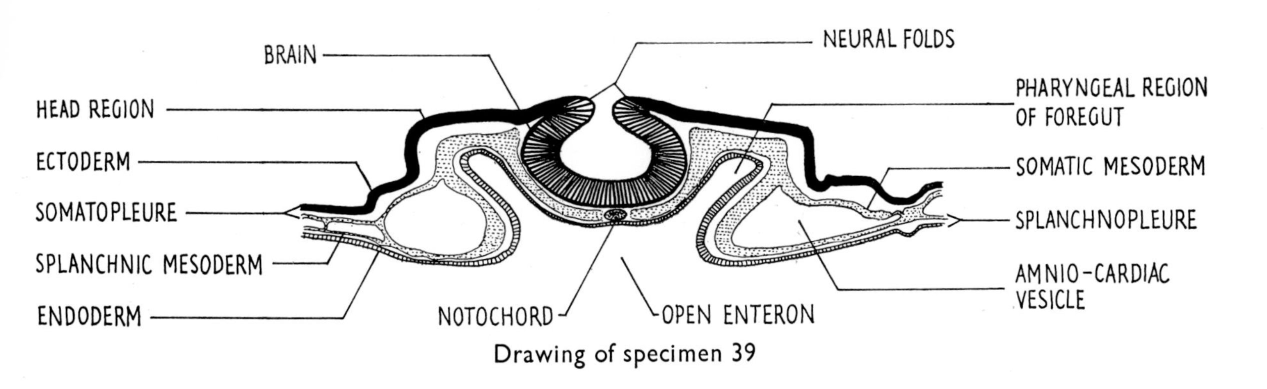
BRAIN
NEURAL FOLDS
HEAD REGION
PHARYNGEAL REGION OF FOREGUT
ECTODERM
SOMATIC MESODERM
SOMATOPLEURE
SPLANCHNOPLEURE
SPLANCHNIC MESODERM
AMNIO-CARDIAC VESICLE
ENDODERM
NOTOCHORD
OPEN ENTERON

Drawing of specimen 39

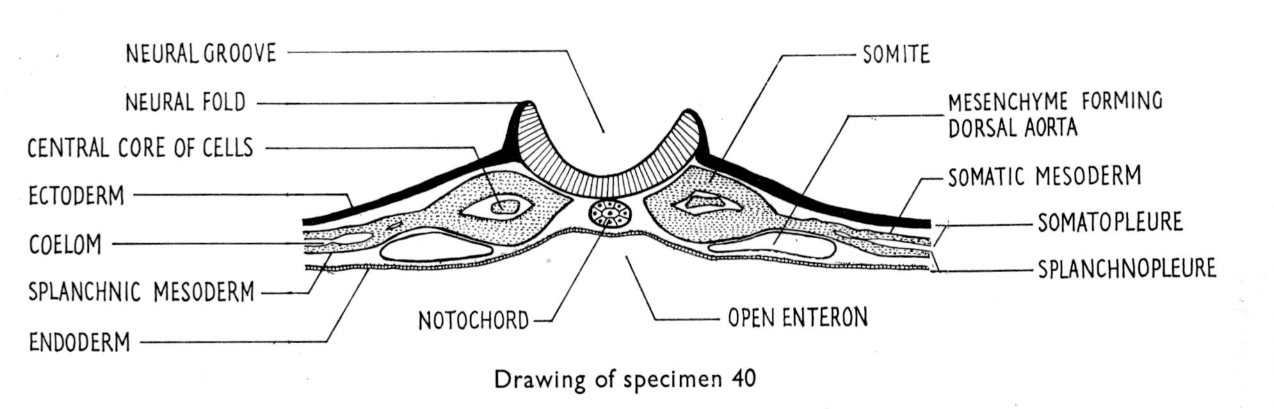
NEURAL GROOVE
SOMITE
NEURAL FOLD
MESENCHYME FORMING DORSAL AORTA
CENTRAL CORE OF CELLS
ECTODERM
SOMATIC MESODERM
SOMATOPLEURE
COELOM
SPLANCHNOPLEURE
SPLANCHNIC MESODERM
NOTOCHORD
OPEN ENTERON
ENDODERM

Drawing of specimen 40

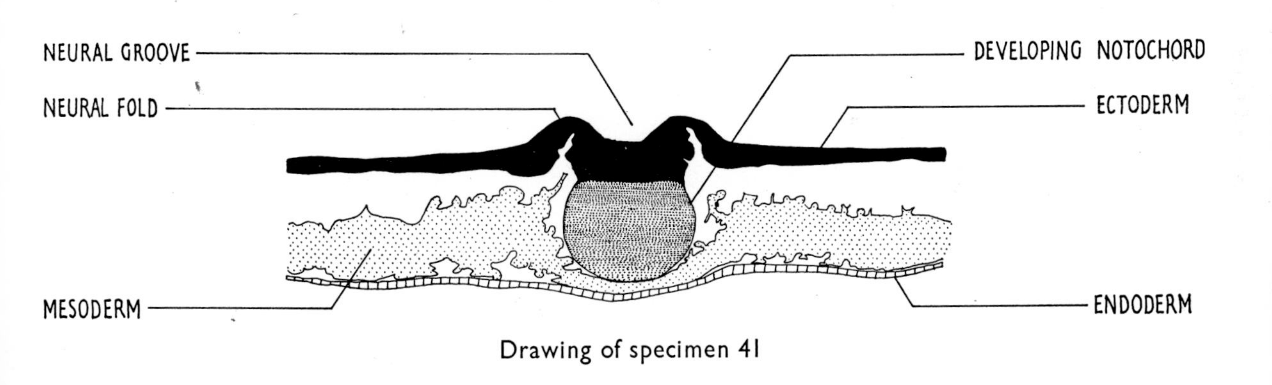
NEURAL GROOVE
DEVELOPING NOTOCHORD
NEURAL FOLD
ECTODERM
MESODERM
ENDODERM

Drawing of specimen 41

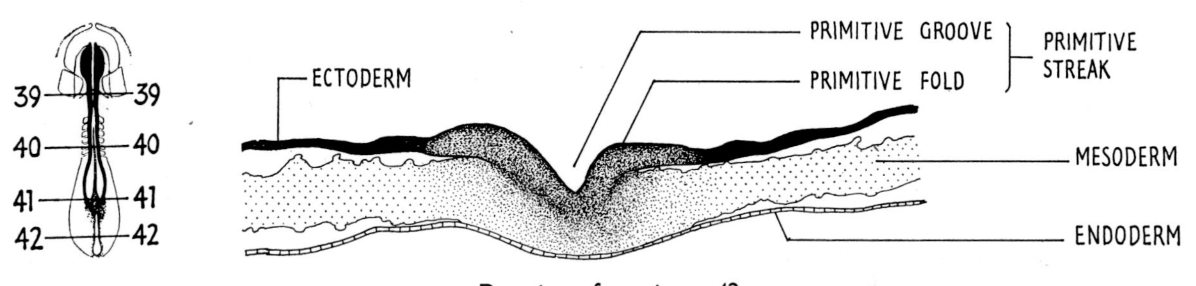
39
39
40
40
41
41
42
42
PRIMITIVE GROOVE
PRIMITIVE FOLD
PRIMITIVE STREAK
ECTODERM
MESODERM
ENDODERM

Drawing of specimen 42

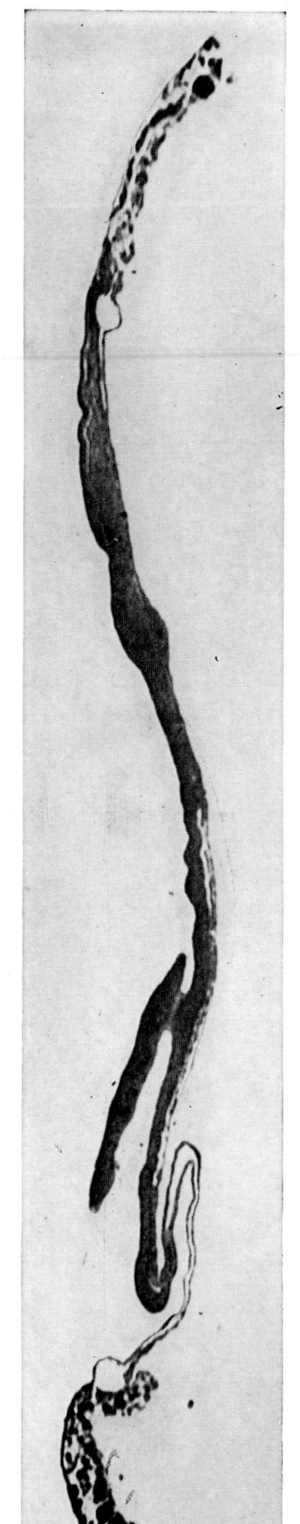

43. **Chick:** 6-somite stage, U.L.S. *mag.* 38×

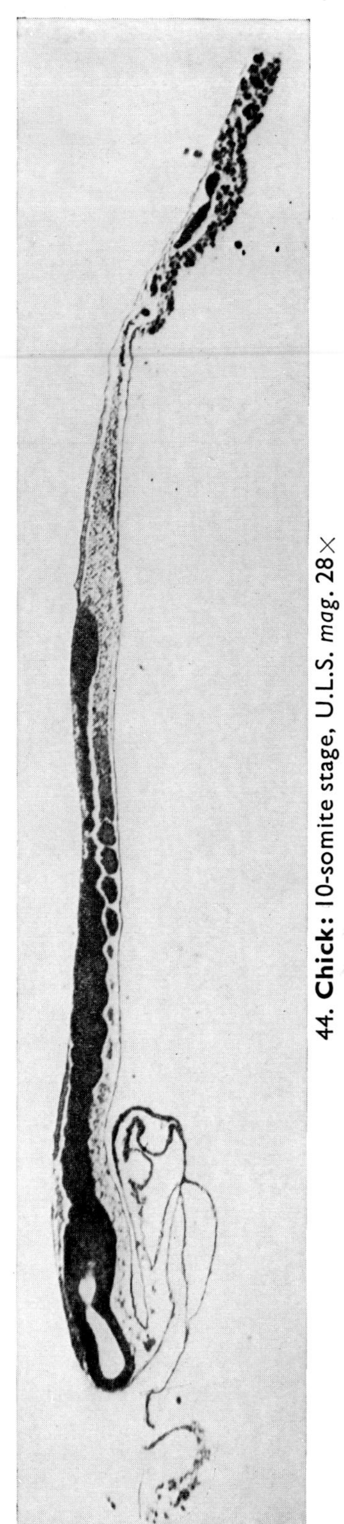

44. **Chick:** 10-somite stage, U.L.S. *mag.* 28×

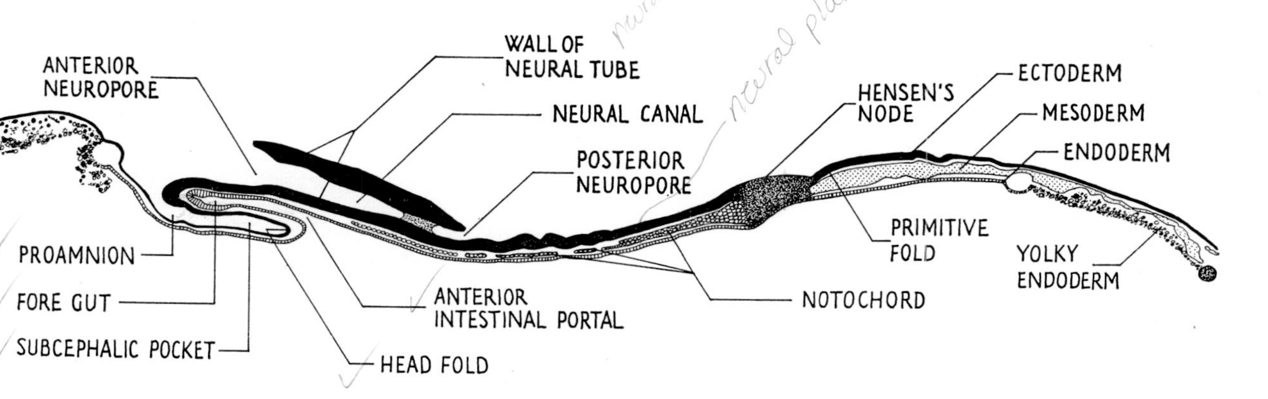
ANTERIOR NEUROPORE
WALL OF NEURAL TUBE
NEURAL CANAL
POSTERIOR NEUROPORE
HENSEN'S NODE
ECTODERM
MESODERM
ENDODERM
PRIMITIVE FOLD
YOLKY ENDODERM
NOTOCHORD
PROAMNION
FORE GUT
SUBCEPHALIC POCKET
ANTERIOR INTESTINAL PORTAL
HEAD FOLD
neural folds?
neural plate

Drawing of specimen 43

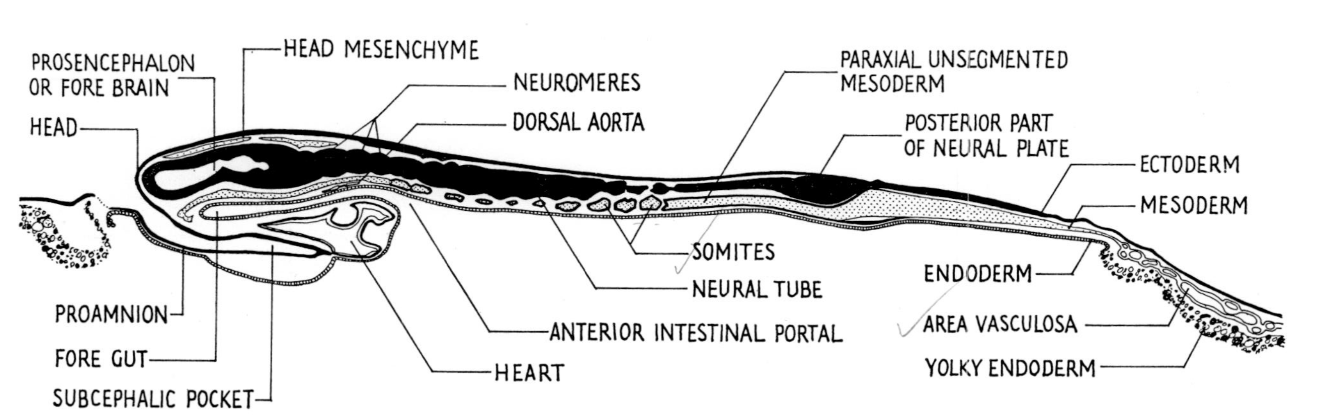
PROSENCEPHALON OR FORE BRAIN
HEAD
HEAD MESENCHYME
NEUROMERES
DORSAL AORTA
PARAXIAL UNSEGMENTED MESODERM
POSTERIOR PART OF NEURAL PLATE
ECTODERM
MESODERM
SOMITES
NEURAL TUBE
ENDODERM
AREA VASCULOSA
YOLKY ENDODERM
ANTERIOR INTESTINAL PORTAL
HEART
PROAMNION
FORE GUT
SUBCEPHALIC POCKET

Drawing of specimen 44

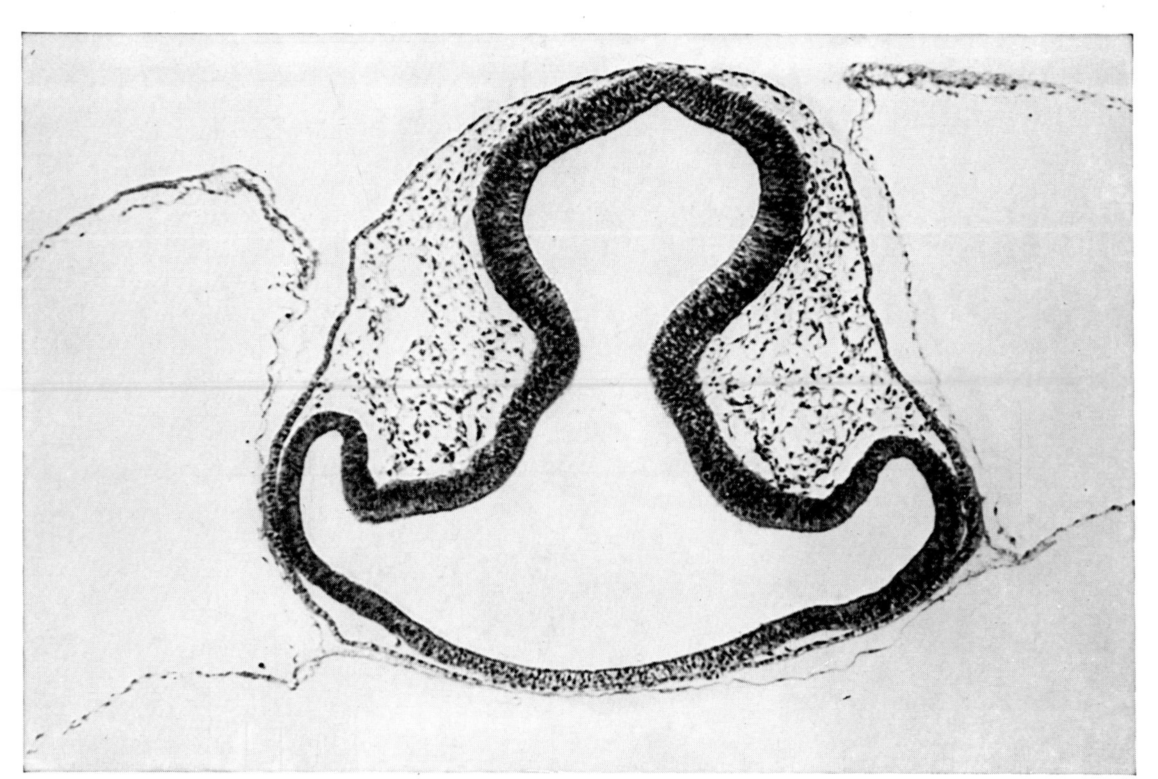

45. **Chick:** 10-somite stage, forebrain region, T.S. *mag. 100×*

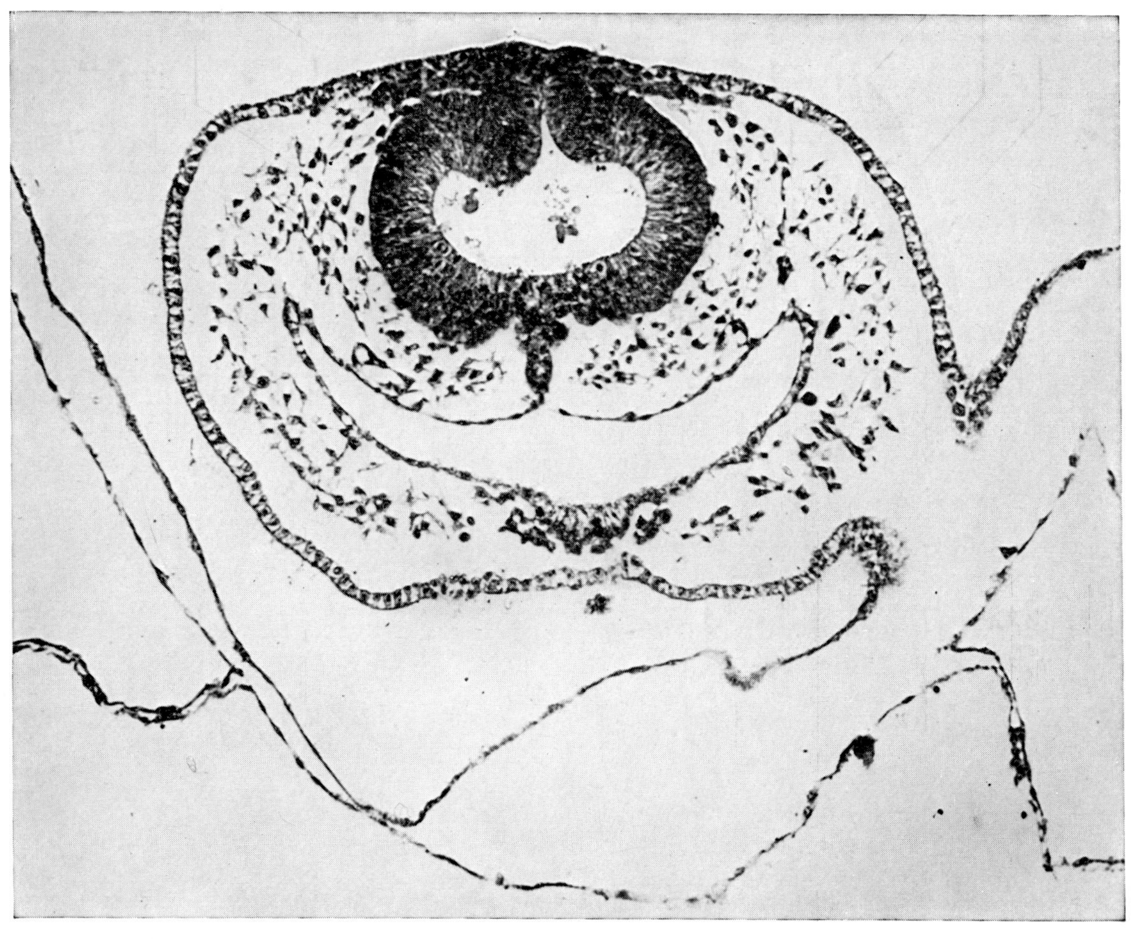

46. **Chick:** 10-somite stage, hindbrain region, T.S. *mag. 200×*

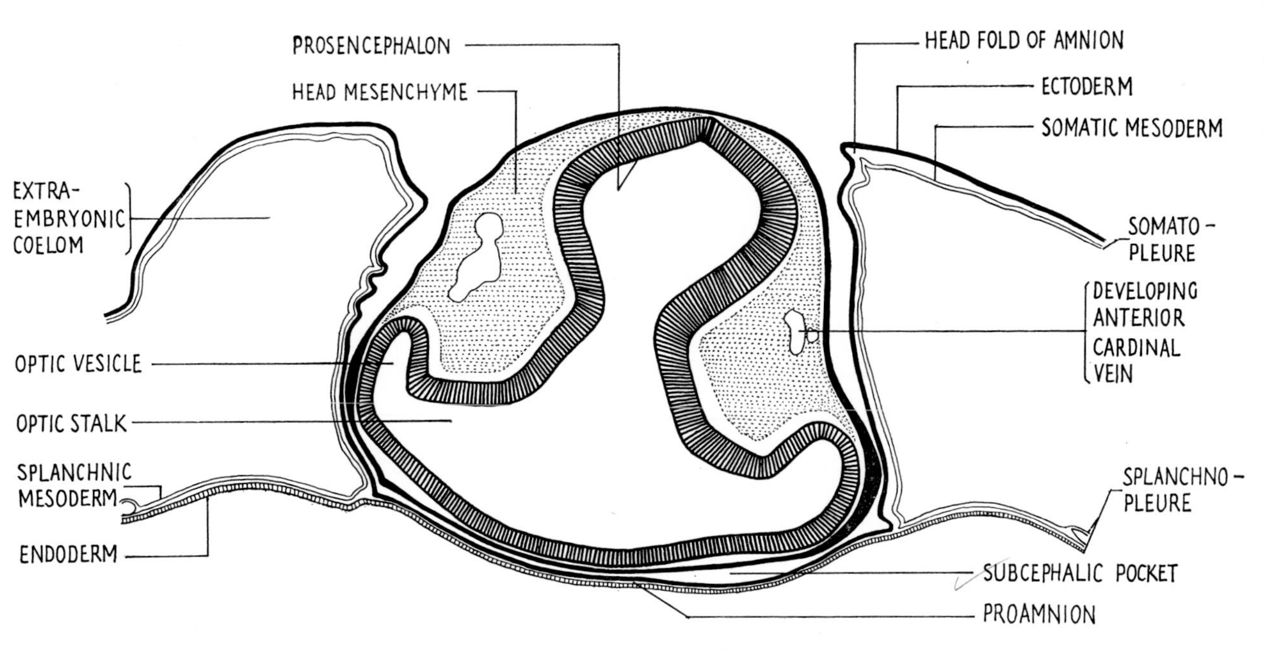
PROSENCEPHALON
HEAD MESENCHYME
EXTRA-EMBRYONIC COELOM
OPTIC VESICLE
OPTIC STALK
SPLANCHNIC MESODERM
ENDODERM
HEAD FOLD OF AMNION
ECTODERM
SOMATIC MESODERM
SOMATO-PLEURE
DEVELOPING ANTERIOR CARDINAL VEIN
SPLANCHNO-PLEURE
SUBCEPHALIC POCKET
PROAMNION

Drawing of specimen 45

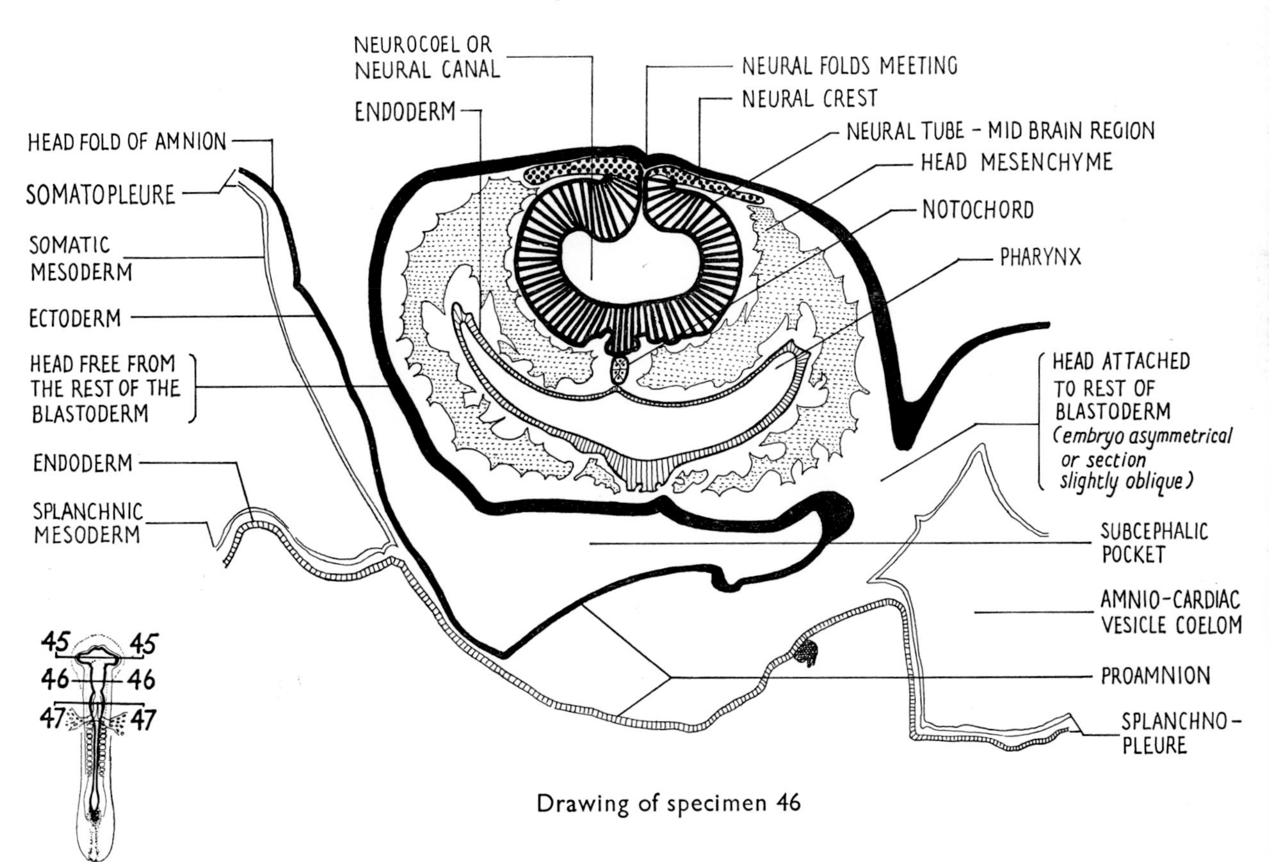
NEUROCOEL OR NEURAL CANAL
ENDODERM
HEAD FOLD OF AMNION
SOMATOPLEURE
SOMATIC MESODERM
ECTODERM
HEAD FREE FROM THE REST OF THE BLASTODERM
ENDODERM
SPLANCHNIC MESODERM
NEURAL FOLDS MEETING
NEURAL CREST
NEURAL TUBE - MID BRAIN REGION
HEAD MESENCHYME
NOTOCHORD
PHARYNX
HEAD ATTACHED TO REST OF BLASTODERM
(embryo asymmetrical or section slightly oblique)
SUBCEPHALIC POCKET
AMNIO-CARDIAC VESICLE COELOM
PROAMNION
SPLANCHNO-PLEURE
45
46
47

Drawing of specimen 46

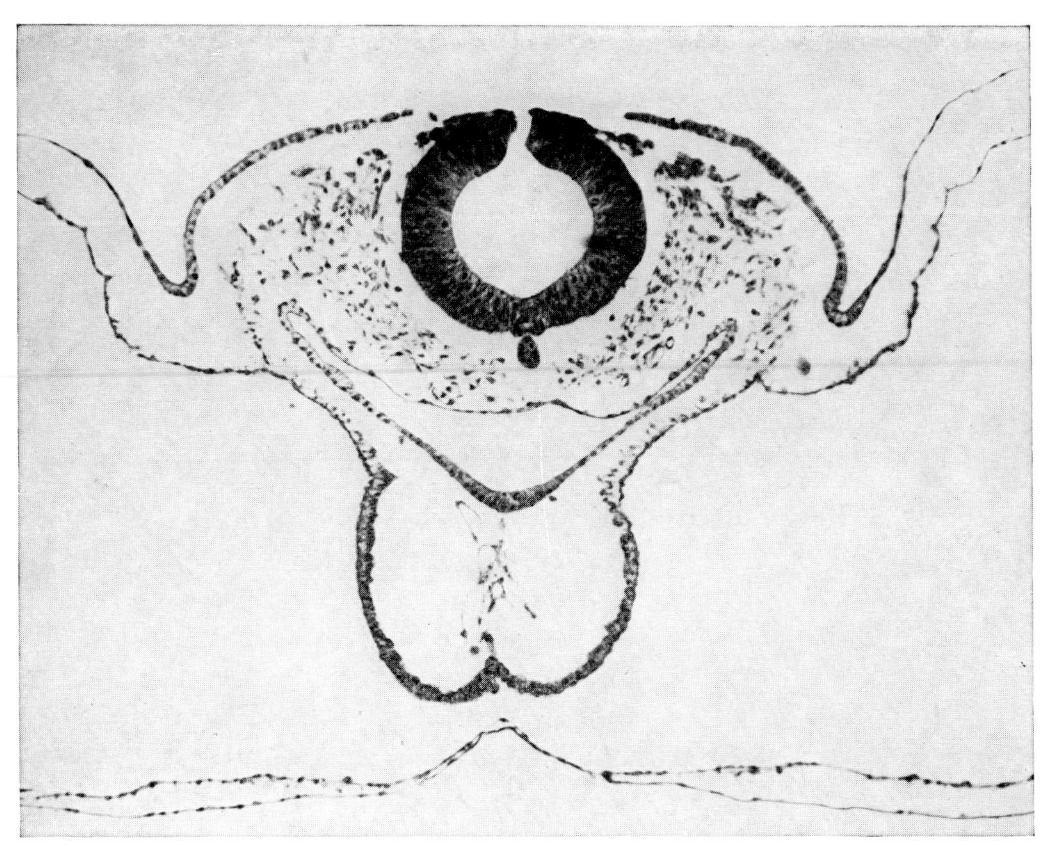

47. **Chick:** 10-somite stage, heart region, T.S. *mag. 150×*

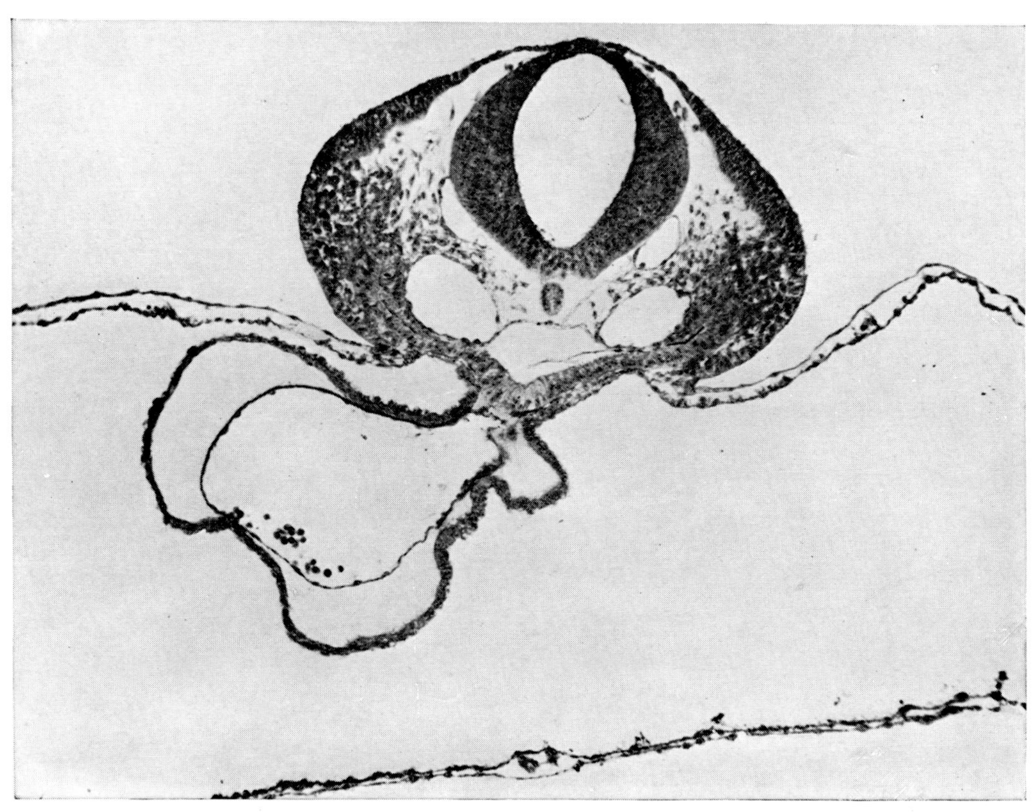

48. **Chick:** 13-somite stage, heart region, T.S. *mag. 150×*

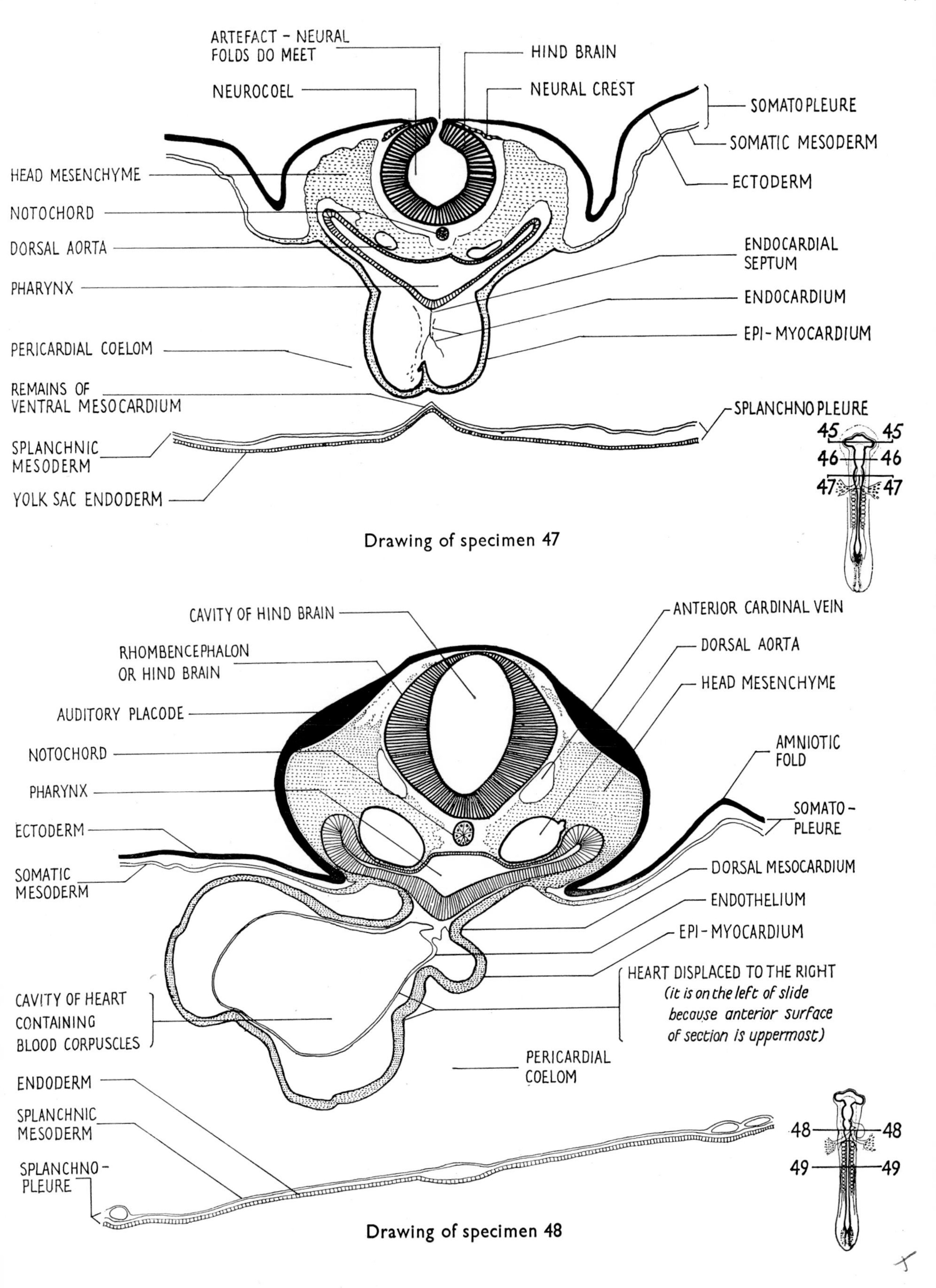
ARTEFACT - NEURAL FOLDS DO MEET
HIND BRAIN
NEUROCOEL
NEURAL CREST
SOMATOPLEURE
SOMATIC MESODERM
ECTODERM
HEAD MESENCHYME
NOTOCHORD
DORSAL AORTA
ENDOCARDIAL SEPTUM
PHARYNX
ENDOCARDIUM
EPI-MYOCARDIUM
PERICARDIAL COELOM
REMAINS OF VENTRAL MESOCARDIUM
SPLANCHNOPLEURE
SPLANCHNIC MESODERM
YOLK SAC ENDODERM
45
46
47

Drawing of specimen 47

CAVITY OF HIND BRAIN
ANTERIOR CARDINAL VEIN
RHOMBENCEPHALON OR HIND BRAIN
DORSAL AORTA
HEAD MESENCHYME
AUDITORY PLACODE
NOTOCHORD
AMNIOTIC FOLD
PHARYNX
SOMATO-PLEURE
ECTODERM
SOMATIC MESODERM
DORSAL MESOCARDIUM
ENDOTHELIUM
EPI-MYOCARDIUM
HEART DISPLACED TO THE RIGHT (it is on the left of slide because anterior surface of section is uppermost)
CAVITY OF HEART CONTAINING BLOOD CORPUSCLES
PERICARDIAL COELOM
ENDODERM
SPLANCHNIC MESODERM
SPLANCHNO-PLEURE
48
49

Drawing of specimen 48

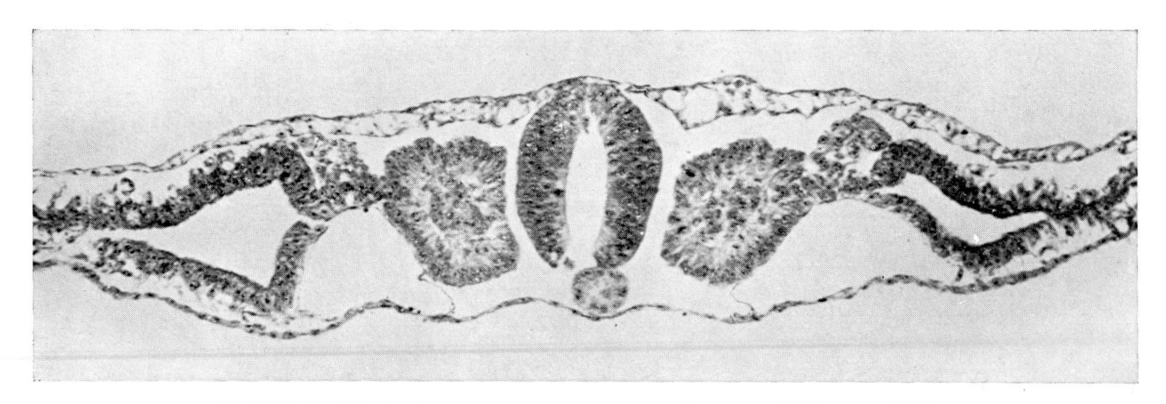

49. **Chick:** 13-somite stage, posterior trunk region, T.S. *mag. 175×*

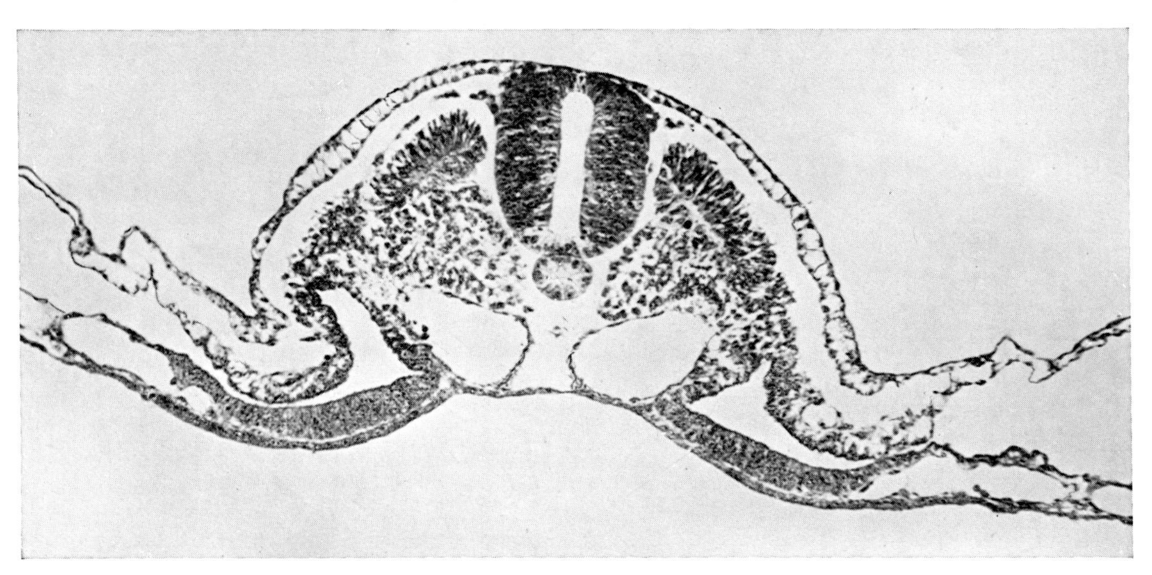

50. **Chick:** 17-somite stage, trunk region, T.S. *mag. 150×*

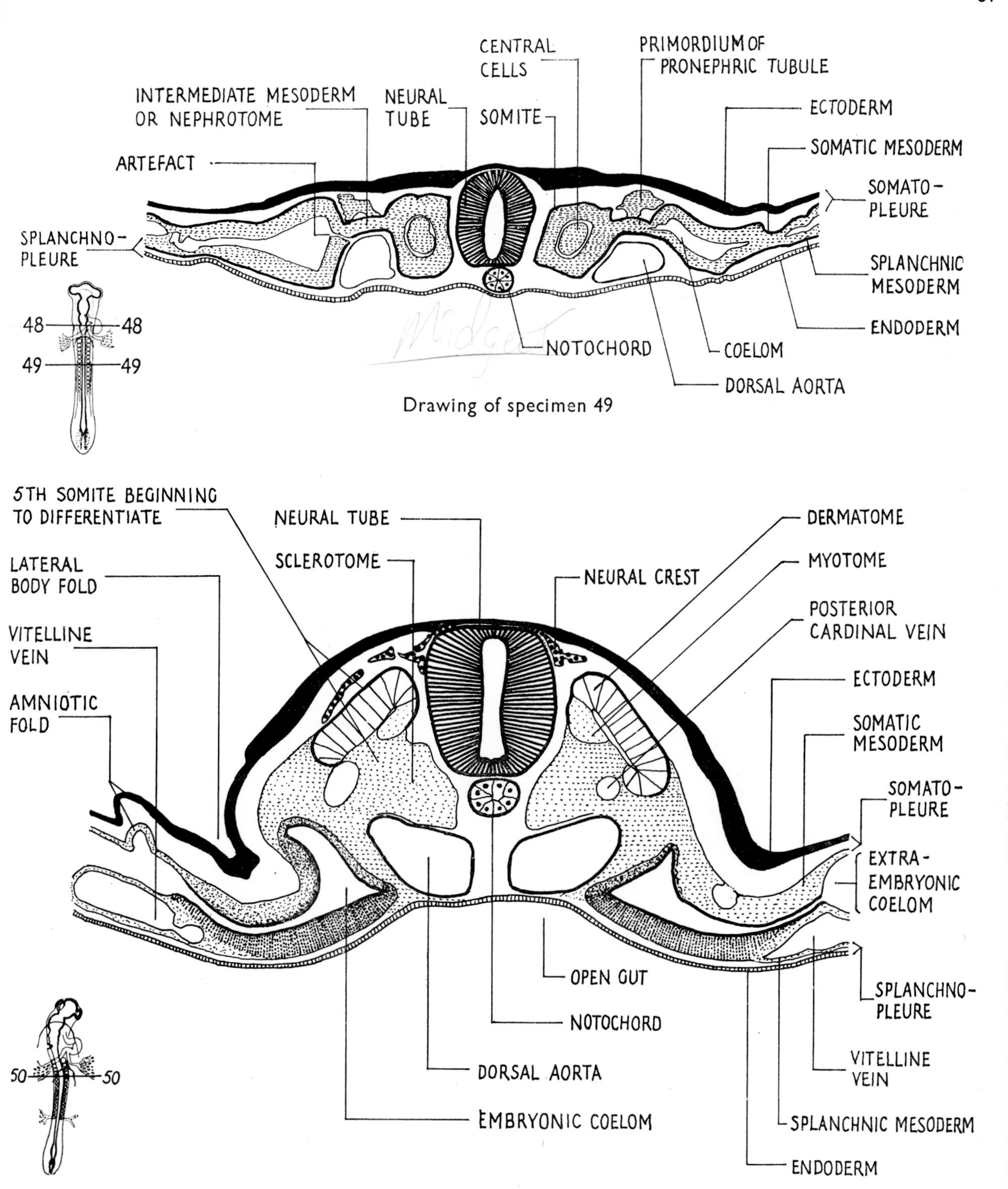
CENTRAL CELLS
PRIMORDIUM OF PRONEPHRIC TUBULE
INTERMEDIATE MESODERM OR NEPHROTOME
NEURAL TUBE
SOMITE
ECTODERM
SOMATIC MESODERM
ARTEFACT
SOMATO-PLEURE
SPLANCHNO-PLEURE
SPLANCHNIC MESODERM
48
48
49
49
ENDODERM
NOTOCHORD
COELOM
DORSAL AORTA

Drawing of specimen 49

5TH SOMITE BEGINNING TO DIFFERENTIATE
NEURAL TUBE
DERMATOME
LATERAL BODY FOLD
SCLEROTOME
MYOTOME
NEURAL CREST
VITELLINE VEIN
POSTERIOR CARDINAL VEIN
AMNIOTIC FOLD
ECTODERM
SOMATIC MESODERM
SOMATO-PLEURE
EXTRA-EMBRYONIC COELOM
SPLANCHNO-PLEURE
OPEN GUT
NOTOCHORD
VITELLINE VEIN
50
50
DORSAL AORTA
SPLANCHNIC MESODERM
EMBRYONIC COELOM
ENDODERM

Drawing of specimen 50

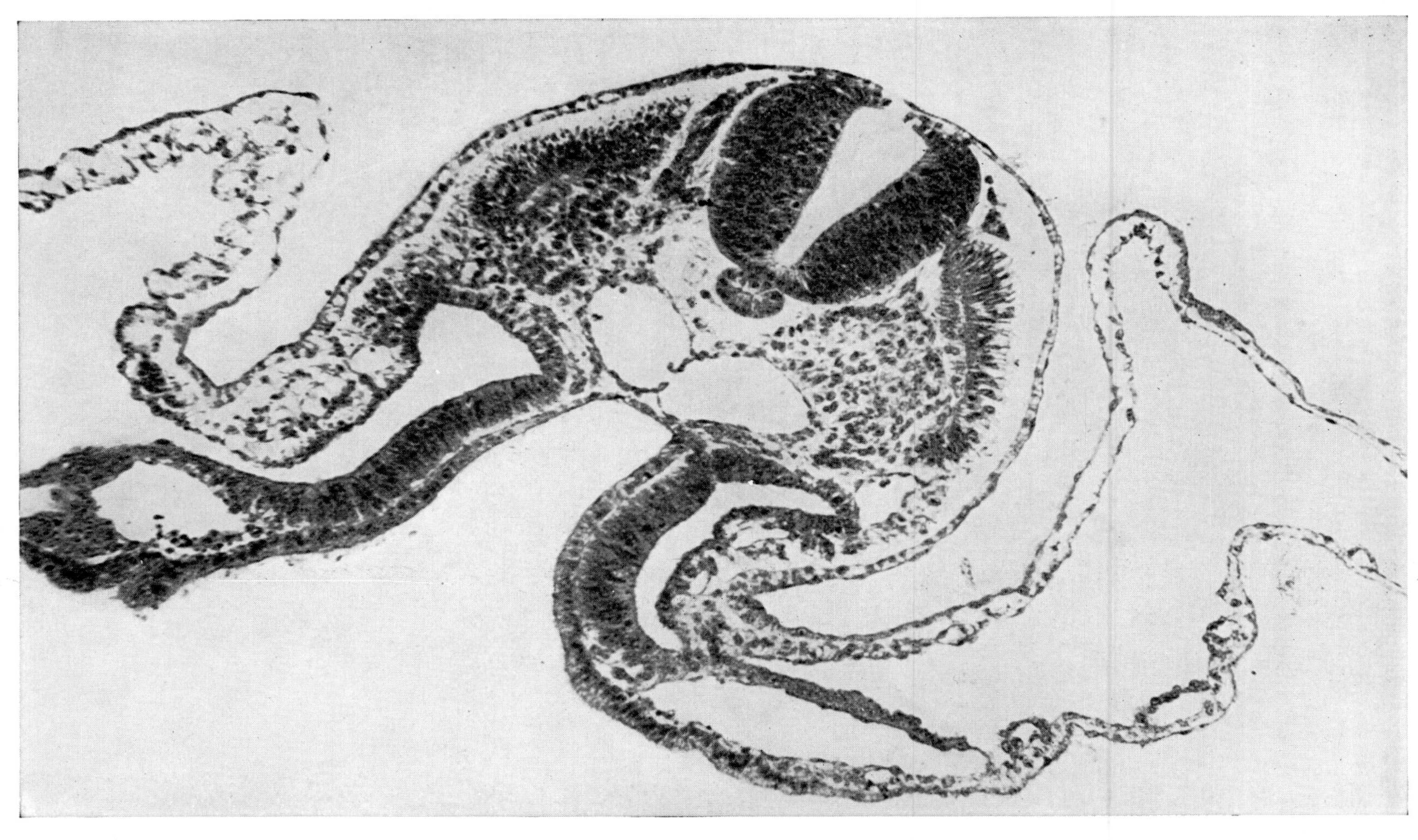

51. **Chick:** 21-somite stage, trunk region, T.S. *mag.* 200×

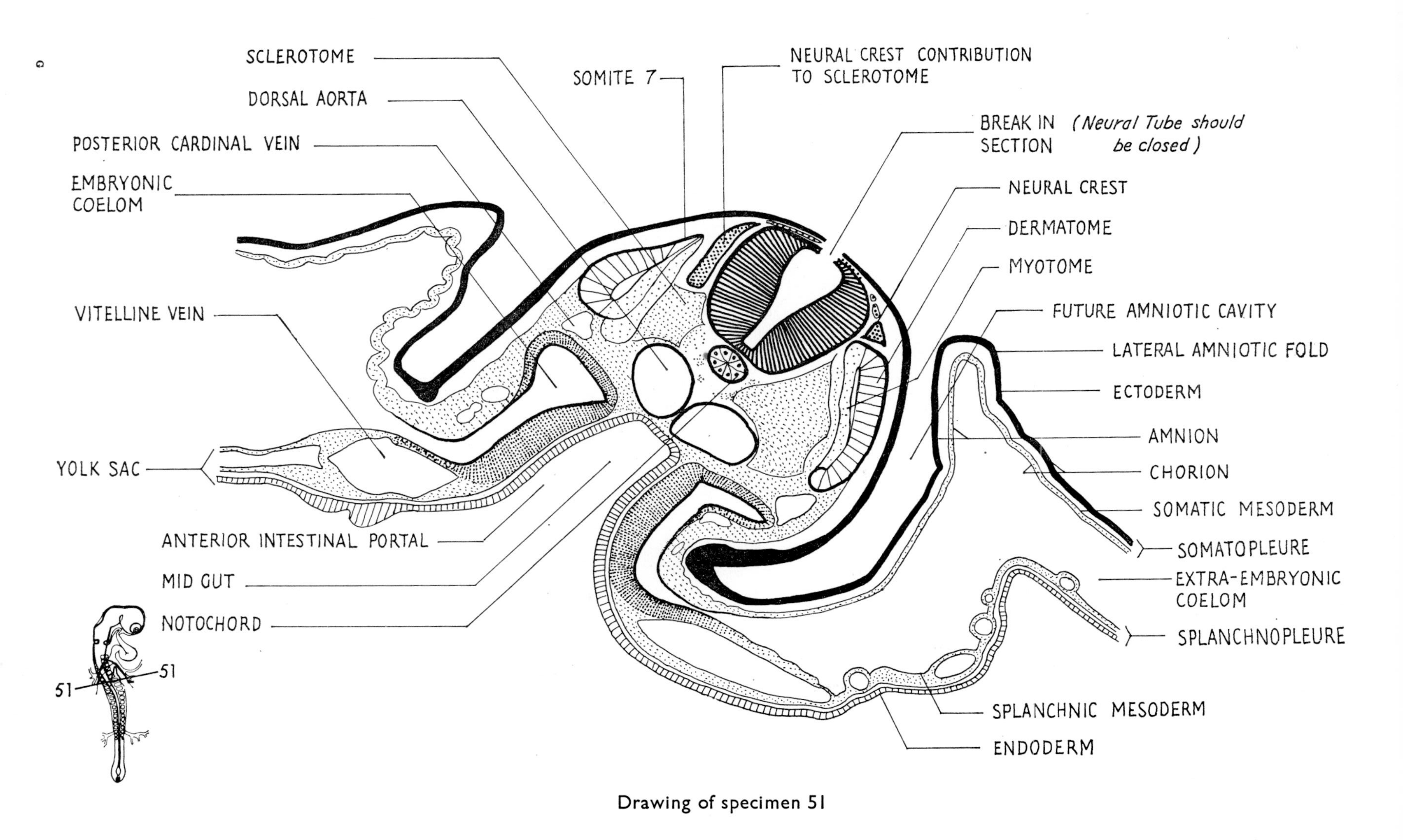
SCLEROTOME
DORSAL AORTA
POSTERIOR CARDINAL VEIN
EMBRYONIC COELOM
SOMITE 7
NEURAL CREST CONTRIBUTION TO SCLEROTOME
BREAK IN SECTION (Neural Tube should be closed)
NEURAL CREST
DERMATOME
MYOTOME
FUTURE AMNIOTIC CAVITY
LATERAL AMNIOTIC FOLD
ECTODERM
AMNION
CHORION
SOMATIC MESODERM
SOMATOPLEURE
EXTRA-EMBRYONIC COELOM
SPLANCHNOPLEURE
SPLANCHNIC MESODERM
ENDODERM
VITELLINE VEIN
YOLK SAC
ANTERIOR INTESTINAL PORTAL
MID GUT
NOTOCHORD
51
51

Drawing of specimen 51

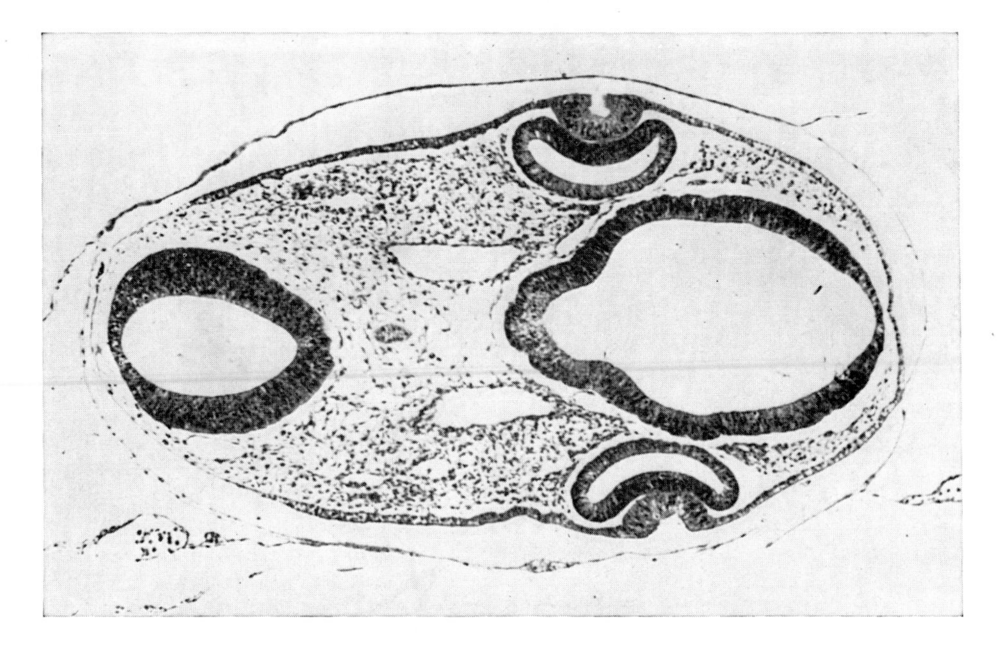

52. **Chick:** 24-somite stage, fore- and hind-brain, T.S. (1). *mag. 45×*

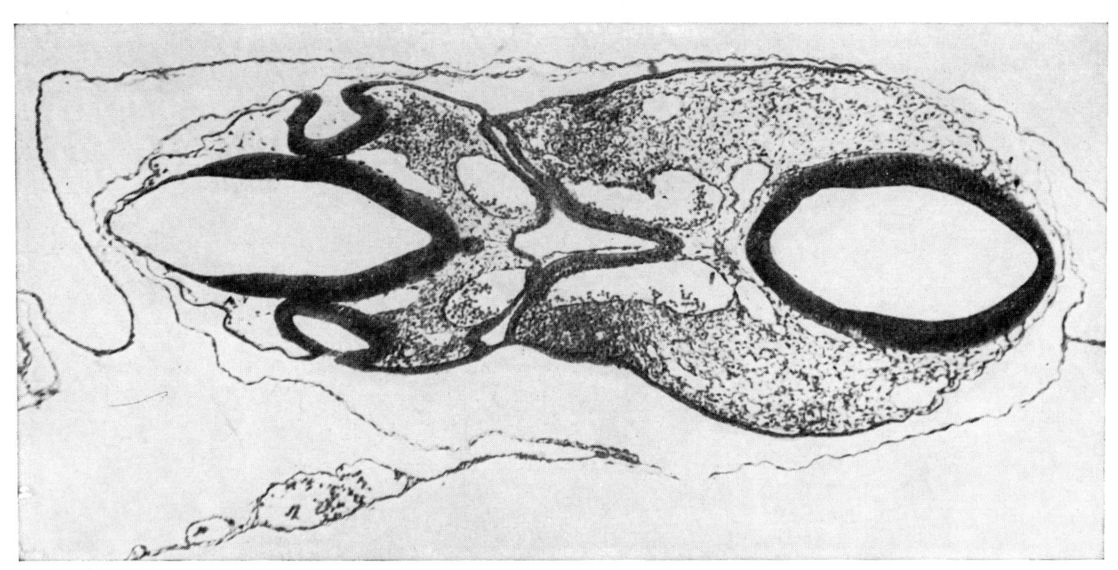

53. **Chick:** 24-somite stage, fore- and hind-brain, T.S. (2). *mag. 70×*

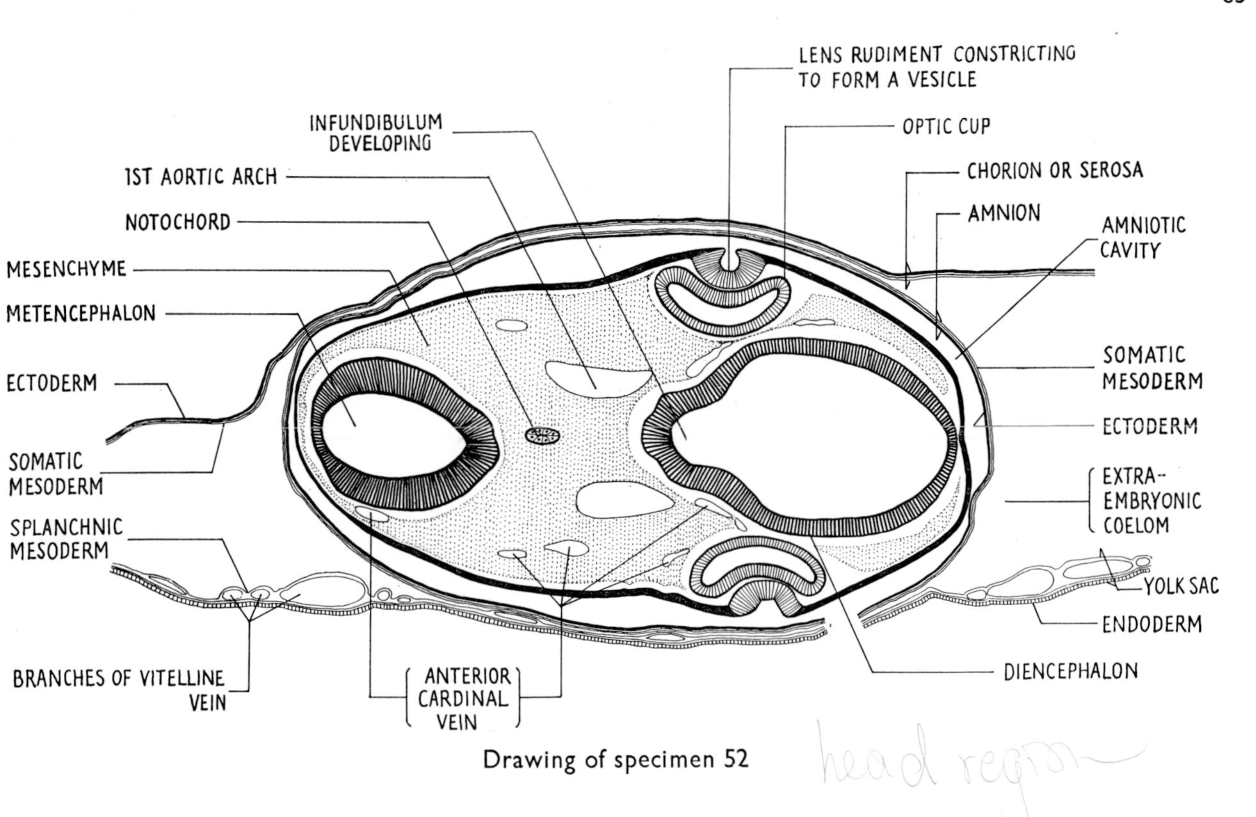
LENS RUDIMENT CONSTRICTING TO FORM A VESICLE
INFUNDIBULUM DEVELOPING
OPTIC CUP
1ST AORTIC ARCH
CHORION OR SEROSA
NOTOCHORD
AMNION
AMNIOTIC CAVITY
MESENCHYME
METENCEPHALON
ECTODERM
SOMATIC MESODERM
SOMATIC MESODERM
ECTODERM
EXTRA-EMBRYONIC COELOM
SPLANCHNIC MESODERM
YOLK SAC
ENDODERM
BRANCHES OF VITELLINE VEIN
ANTERIOR CARDINAL VEIN
DIENCEPHALON

Drawing of specimen 52

head region

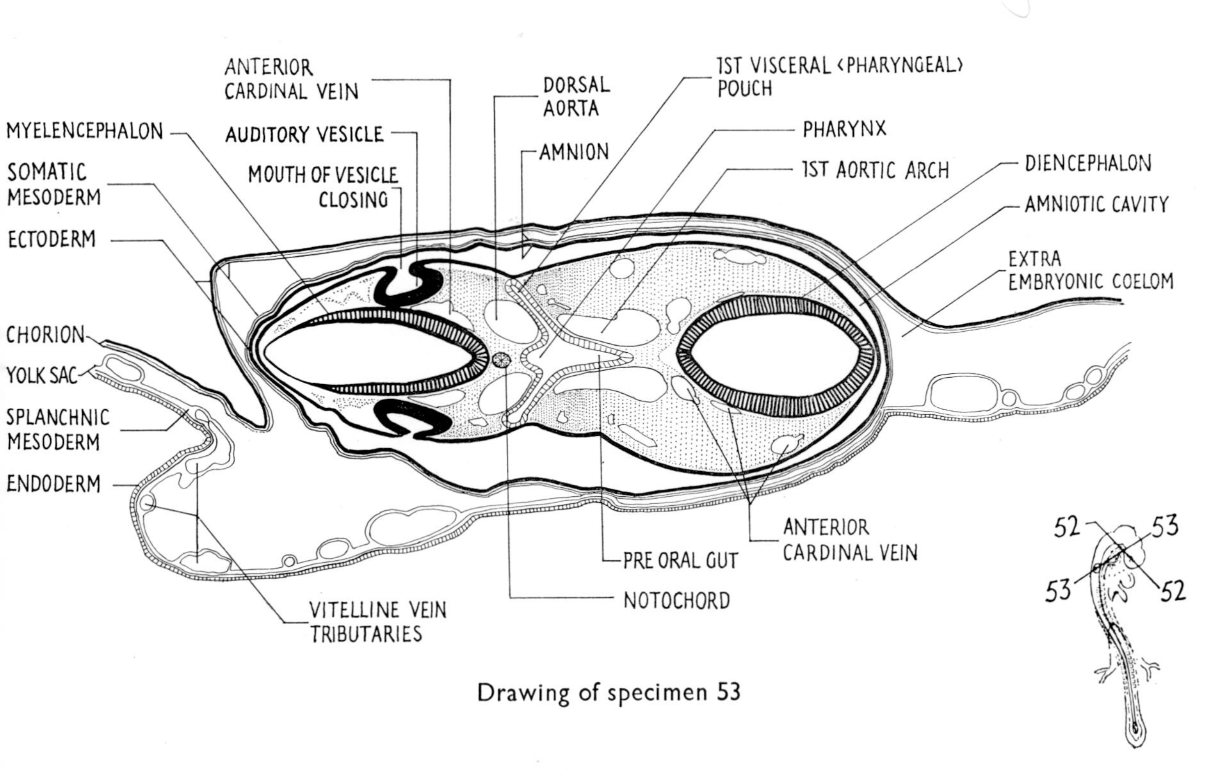
ANTERIOR CARDINAL VEIN
DORSAL AORTA
1ST VISCERAL (PHARYNGEAL) POUCH
MYELENCEPHALON
AUDITORY VESICLE
AMNION
PHARYNX
SOMATIC MESODERM
MOUTH OF VESICLE CLOSING
1ST AORTIC ARCH
DIENCEPHALON
AMNIOTIC CAVITY
ECTODERM
EXTRA EMBRYONIC COELOM
CHORION
YOLK SAC
SPLANCHNIC MESODERM
ENDODERM
ANTERIOR CARDINAL VEIN
PRE ORAL GUT
NOTOCHORD
VITELLINE VEIN TRIBUTARIES
52
53
53
52

Drawing of specimen 53

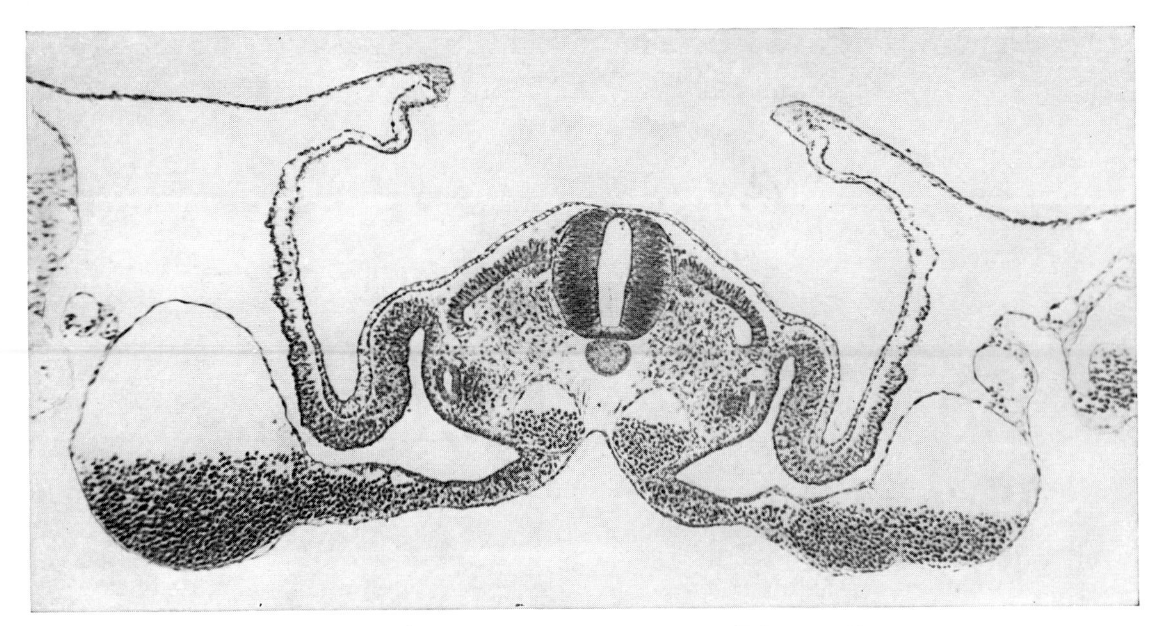

54. **Chick:** 27-somite stage, trunk region, T.S. *mag. 80×*

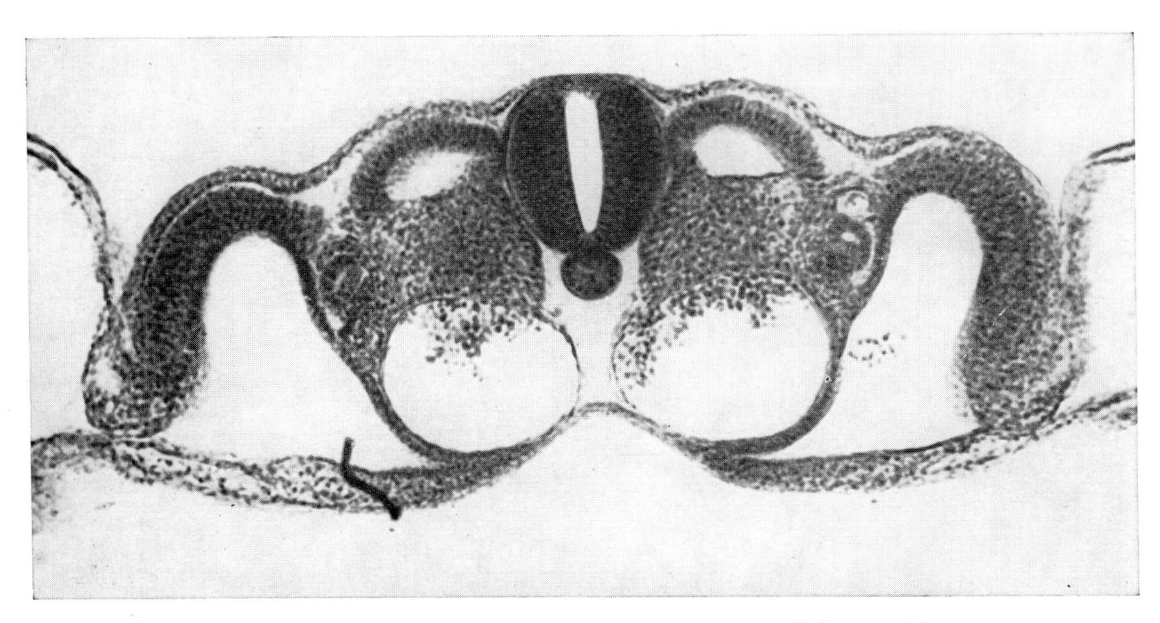

55. **Chick:** 27-somite stage, posterior trunk region, T.S. *mag. 95×*

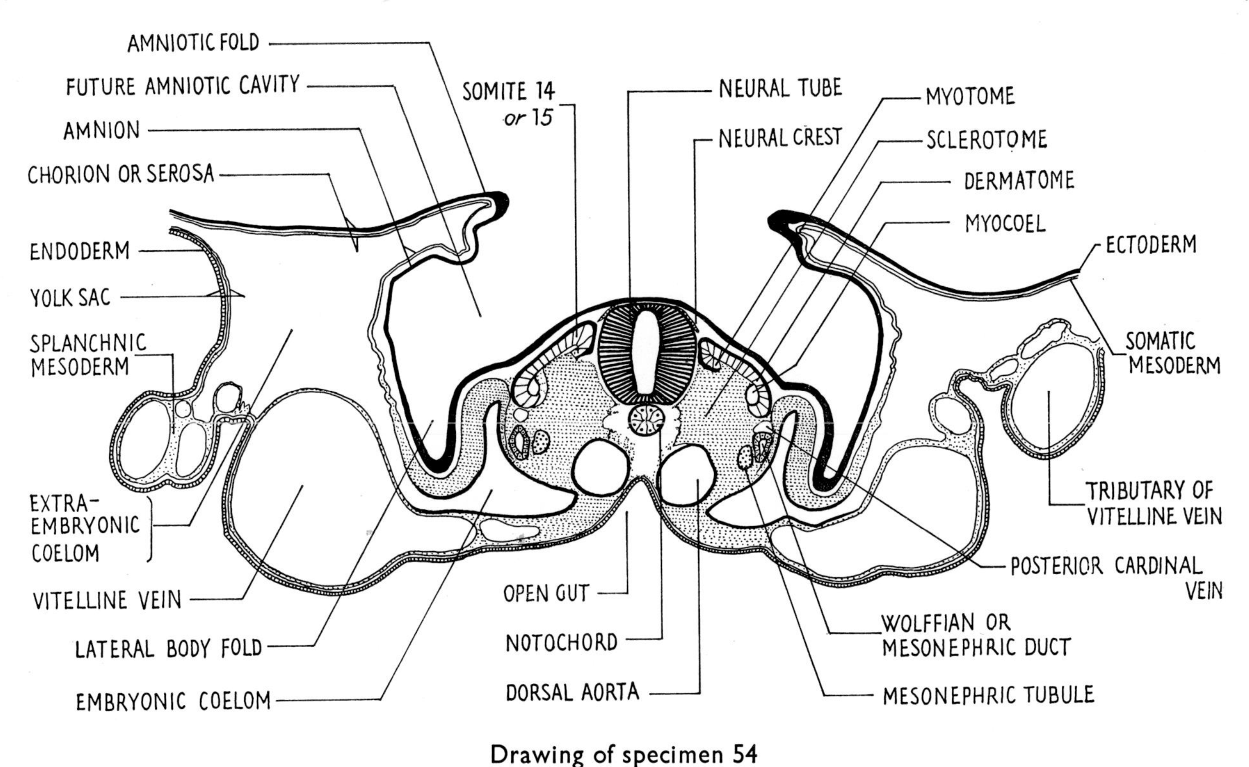
AMNIOTIC FOLD
FUTURE AMNIOTIC CAVITY
AMNION
CHORION OR SEROSA
ENDODERM
YOLK SAC
SPLANCHNIC MESODERM
EXTRA-EMBRYONIC COELOM
VITELLINE VEIN
LATERAL BODY FOLD
EMBRYONIC COELOM
SOMITE 14 or 15
NEURAL TUBE
NEURAL CREST
OPEN GUT
NOTOCHORD
DORSAL AORTA
MYOTOME
SCLEROTOME
DERMATOME
MYOCOEL
ECTODERM
SOMATIC MESODERM
TRIBUTARY OF VITELLINE VEIN
POSTERIOR CARDINAL VEIN
WOLFFIAN OR MESONEPHRIC DUCT
MESONEPHRIC TUBULE

Drawing of specimen 54

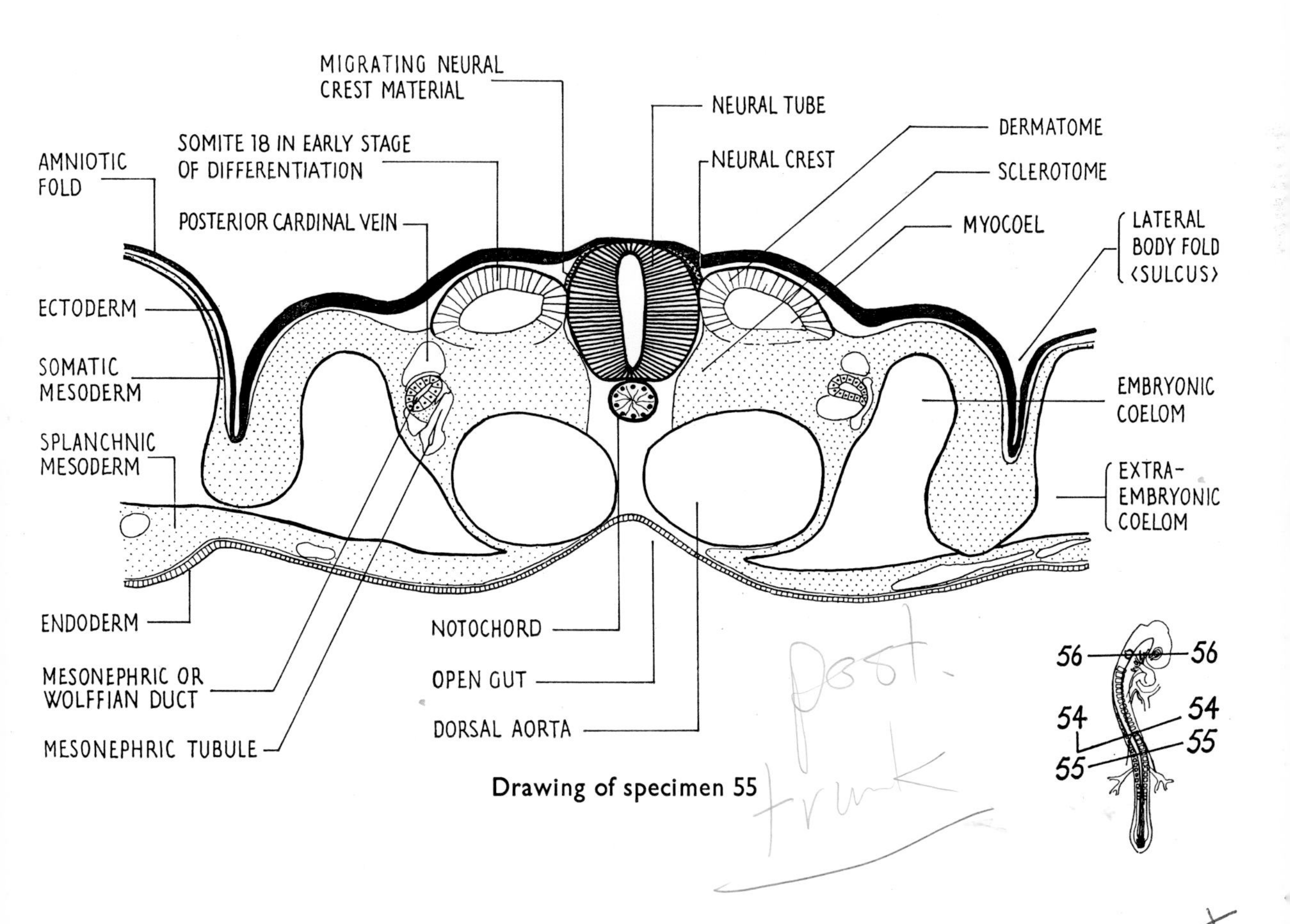
MIGRATING NEURAL CREST MATERIAL
SOMITE 18 IN EARLY STAGE OF DIFFERENTIATION
AMNIOTIC FOLD
POSTERIOR CARDINAL VEIN
ECTODERM
SOMATIC MESODERM
SPLANCHNIC MESODERM
ENDODERM
MESONEPHRIC OR WOLFFIAN DUCT
MESONEPHRIC TUBULE
NEURAL TUBE
NEURAL CREST
NOTOCHORD
OPEN GUT
DORSAL AORTA
DERMATOME
SCLEROTOME
MYOCOEL
LATERAL BODY FOLD (SULCUS)
EMBRYONIC COELOM
EXTRA-EMBRYONIC COELOM
56
56
54
54
55
55
post. trunk

Drawing of specimen 55

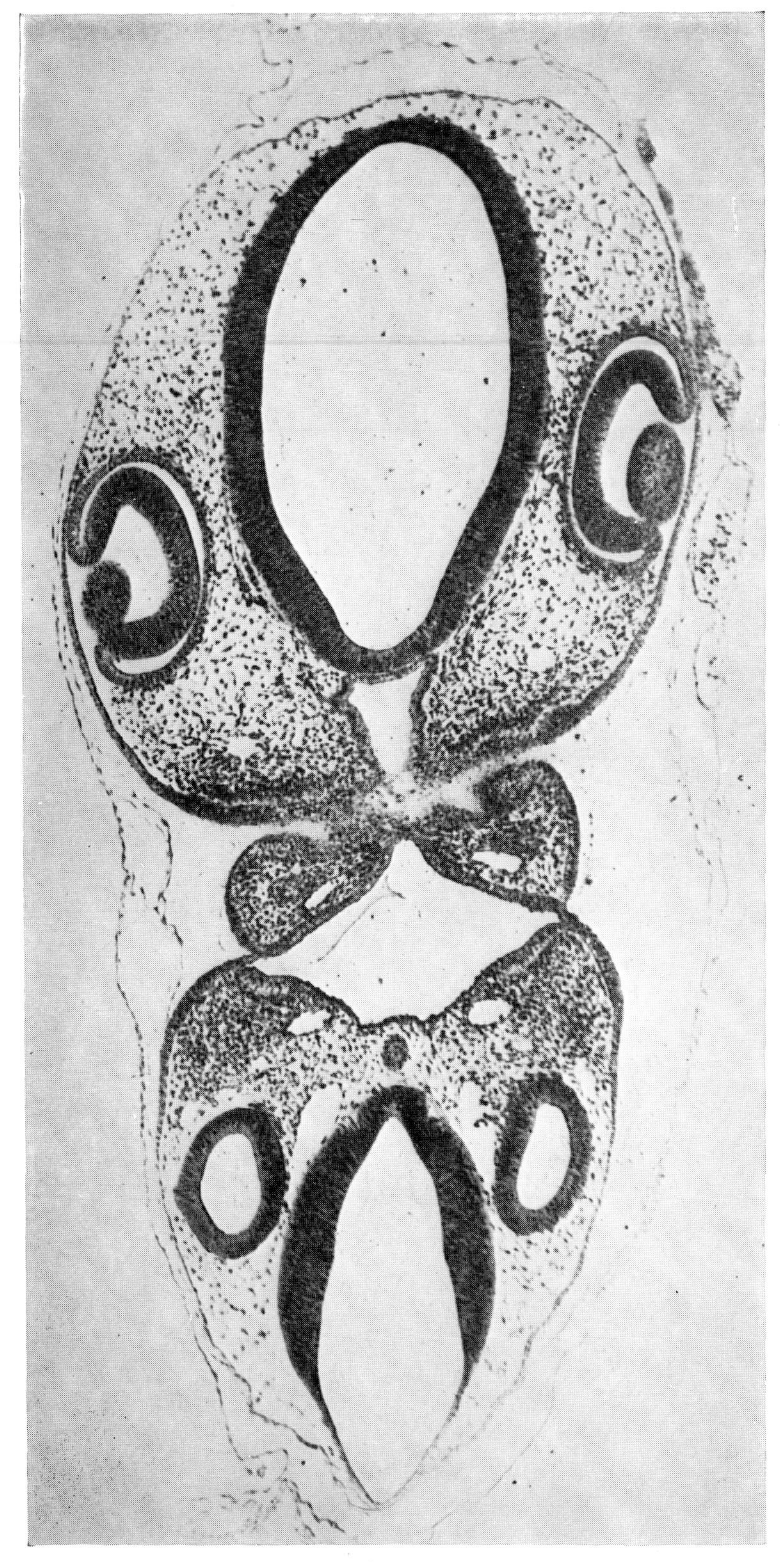

56. **Chick:** 27-somite stage, eye and ear region, T.S. *mag.* *90×*

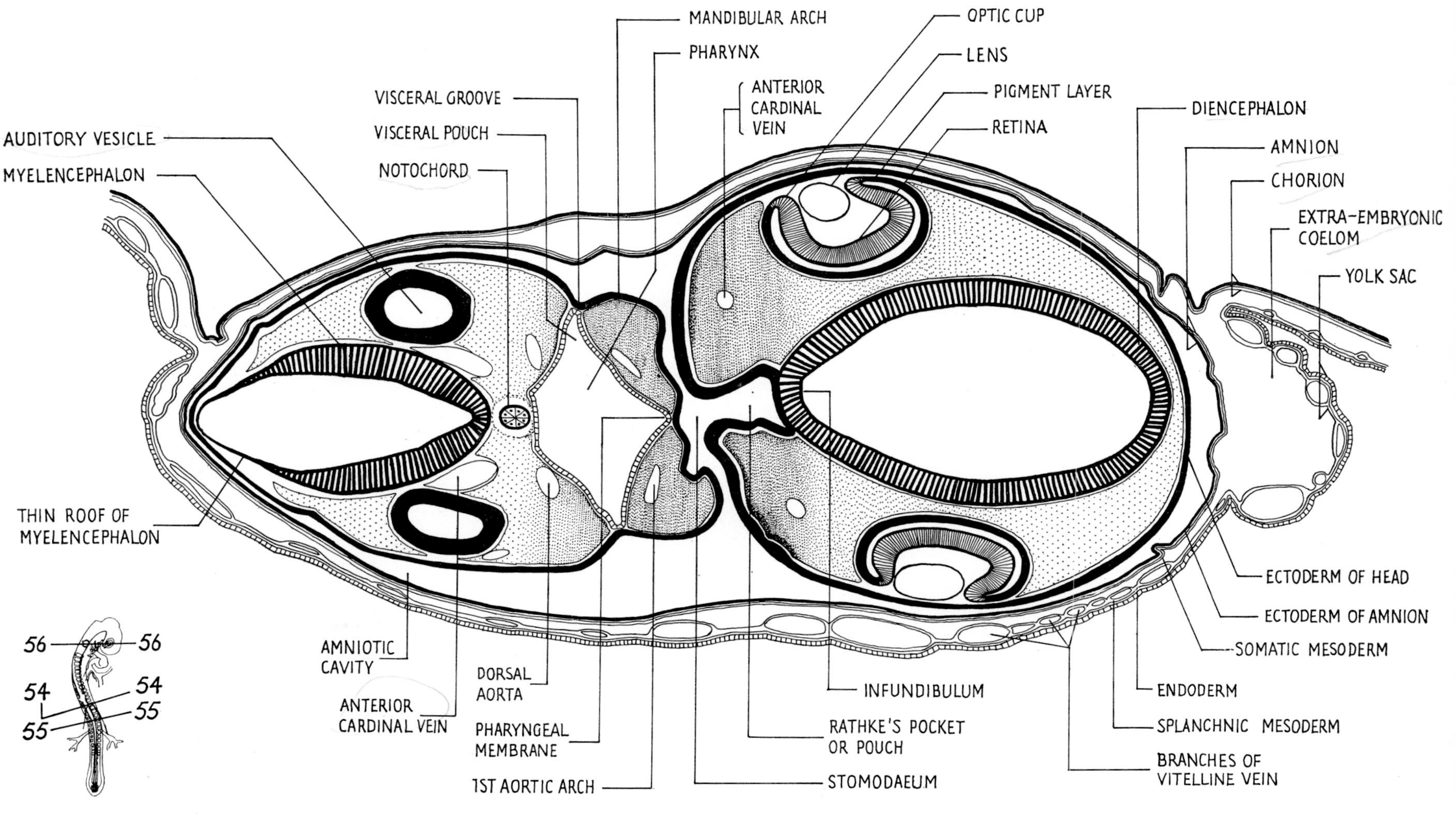
MANDIBULAR ARCH
PHARYNX
OPTIC CUP
LENS
PIGMENT LAYER
RETINA
ANTERIOR CARDINAL VEIN
VISCERAL GROOVE
VISCERAL POUCH
NOTOCHORD
AUDITORY VESICLE
MYELENCEPHALON
DIENCEPHALON
AMNION
CHORION
EXTRA-EMBRYONIC COELOM
YOLK SAC
THIN ROOF OF MYELENCEPHALON
ECTODERM OF HEAD
ECTODERM OF AMNION
SOMATIC MESODERM
ENDODERM
SPLANCHNIC MESODERM
BRANCHES OF VITELLINE VEIN
AMNIOTIC CAVITY
ANTERIOR CARDINAL VEIN
DORSAL AORTA
PHARYNGEAL MEMBRANE
1ST AORTIC ARCH
INFUNDIBULUM
RATHKE'S POCKET OR POUCH
STOMODAEUM
56
56
54
54
55
55

Drawing of specimen 56

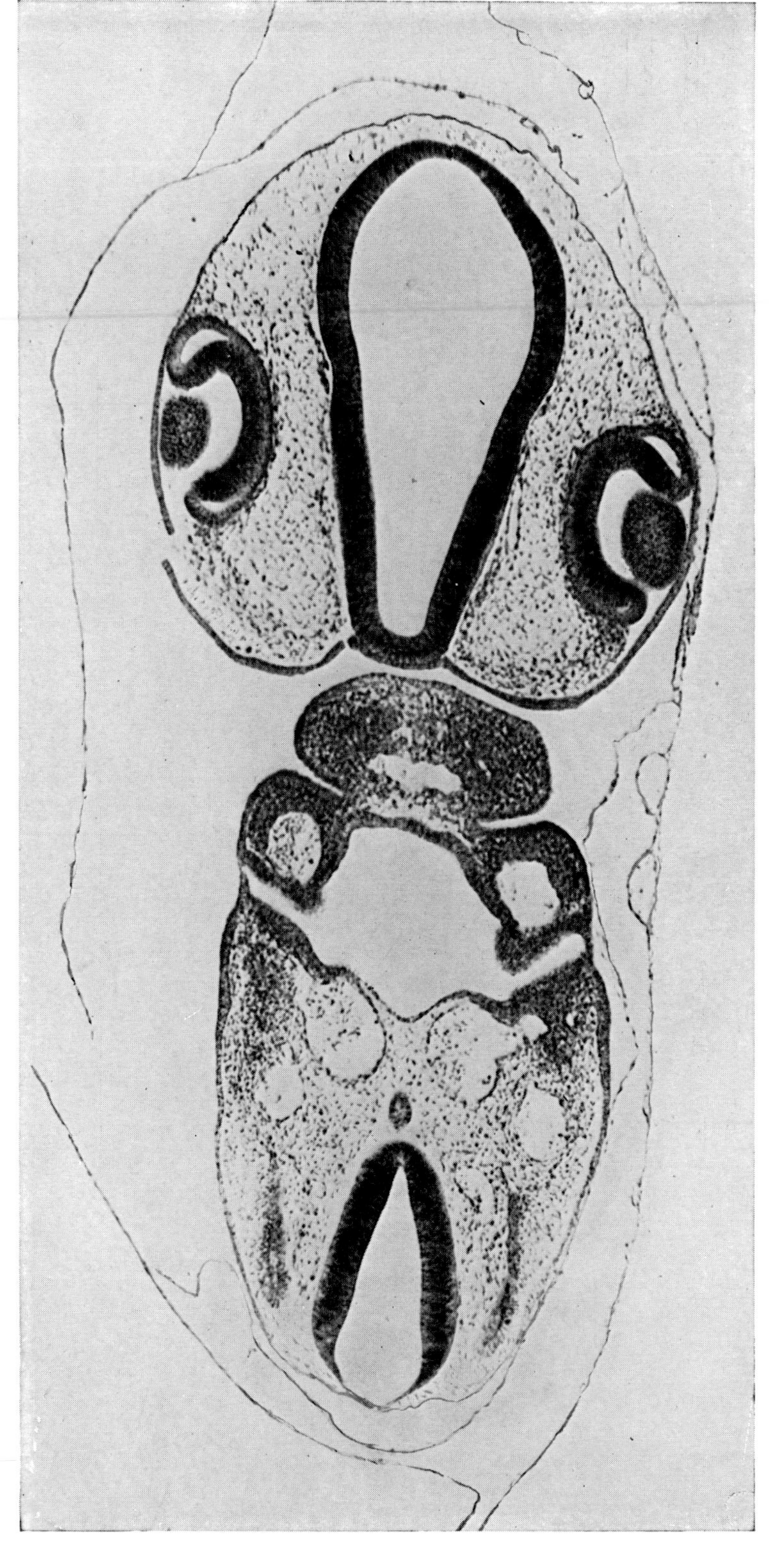

57. Chick: 30-somite stage, fore- and hind-brain, T.S. *mag.* 75×

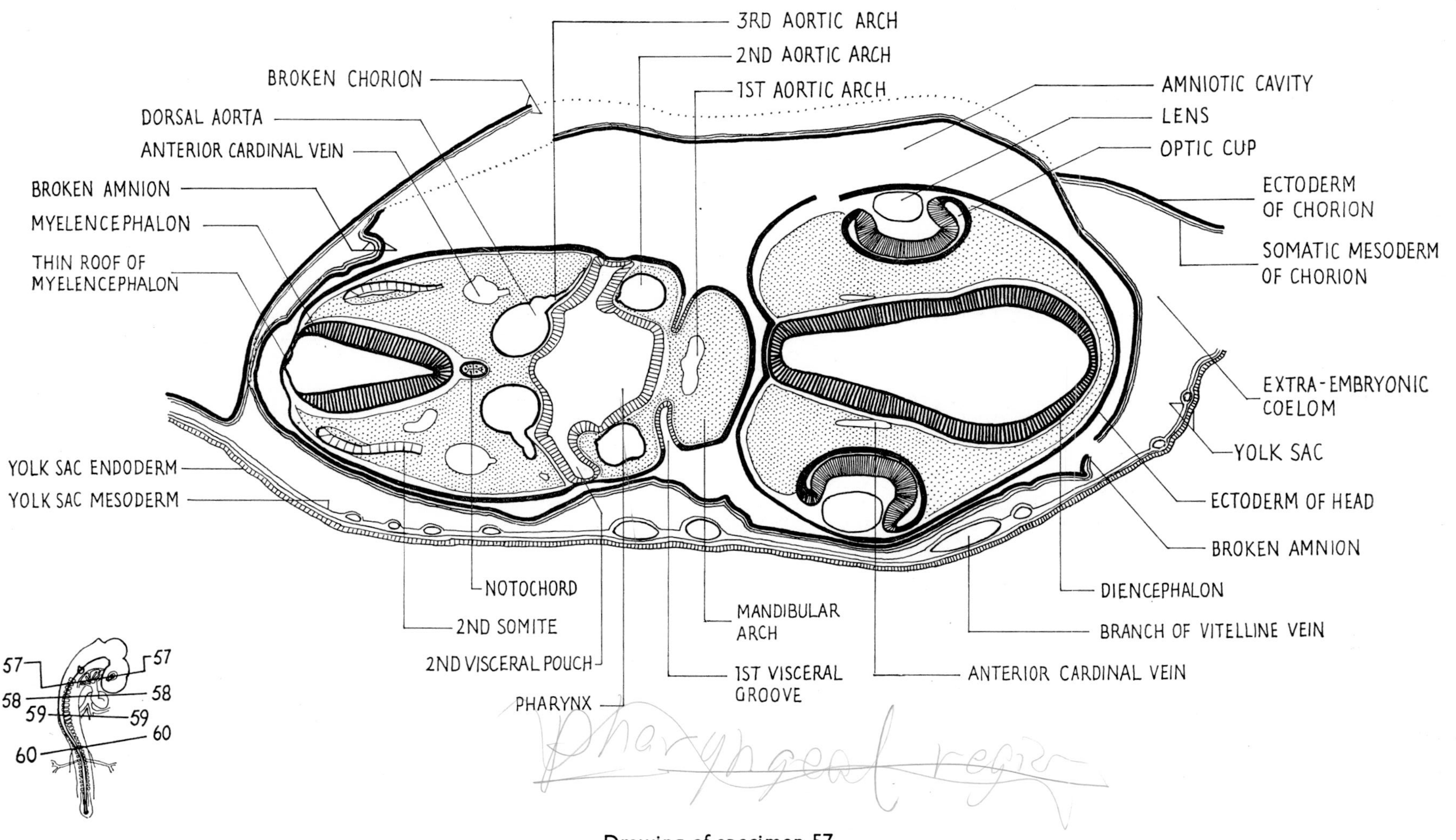
3RD AORTIC ARCH
2ND AORTIC ARCH
1ST AORTIC ARCH
BROKEN CHORION
DORSAL AORTA
ANTERIOR CARDINAL VEIN
BROKEN AMNION
MYELENCEPHALON
THIN ROOF OF MYELENCEPHALON
AMNIOTIC CAVITY
LENS
OPTIC CUP
ECTODERM OF CHORION
SOMATIC MESODERM OF CHORION
EXTRA-EMBRYONIC COELOM
YOLK SAC
ECTODERM OF HEAD
BROKEN AMNION
YOLK SAC ENDODERM
YOLK SAC MESODERM
NOTOCHORD
2ND SOMITE
2ND VISCERAL POUCH
PHARYNX
MANDIBULAR ARCH
1ST VISCERAL GROOVE
DIENCEPHALON
BRANCH OF VITELLINE VEIN
ANTERIOR CARDINAL VEIN
57
57
58
58
59
59
60
60

Drawing of specimen 57

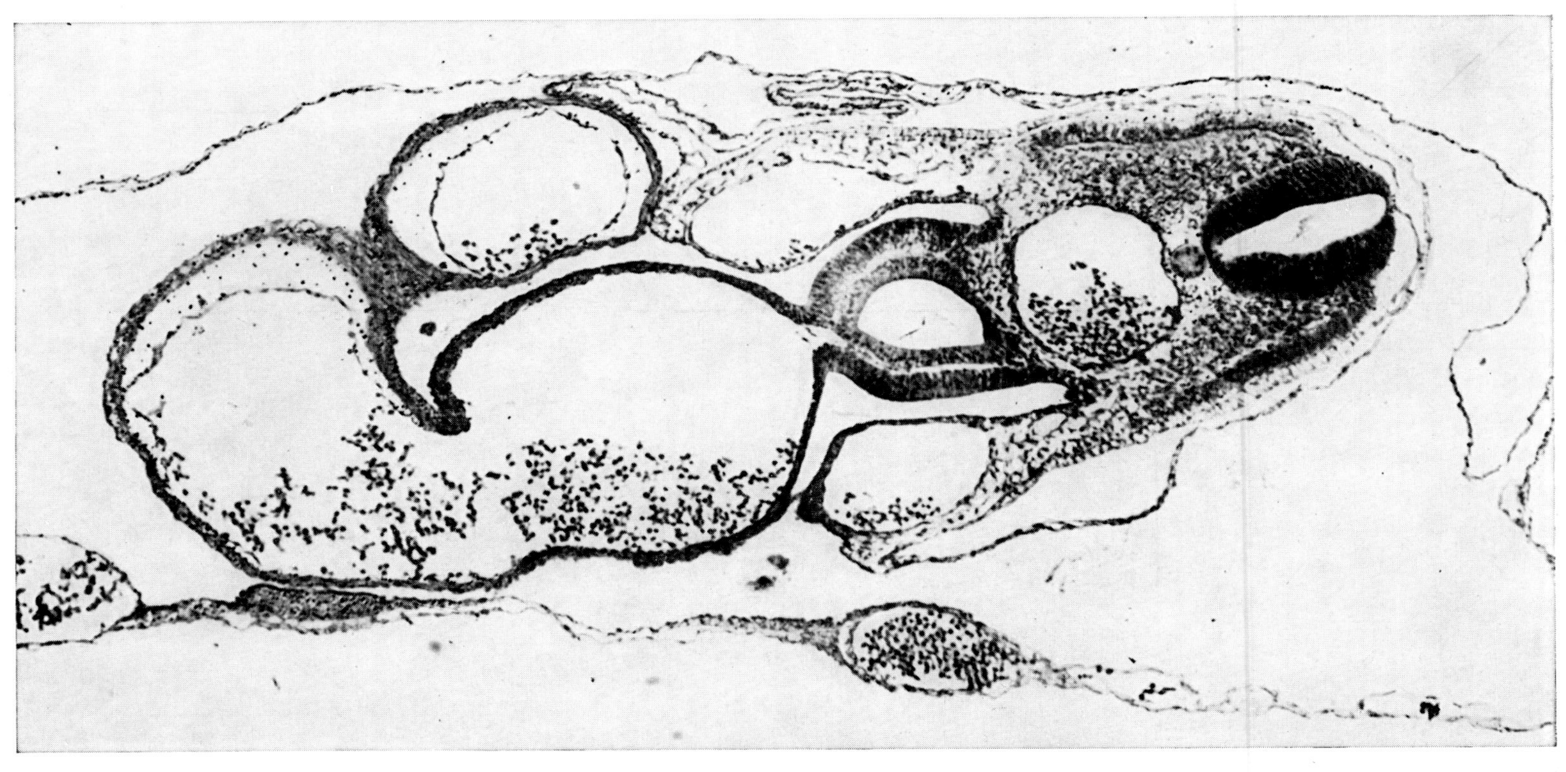

58. **Chick:** 30-somite stage, heart region, T.S. *mag.* 130×

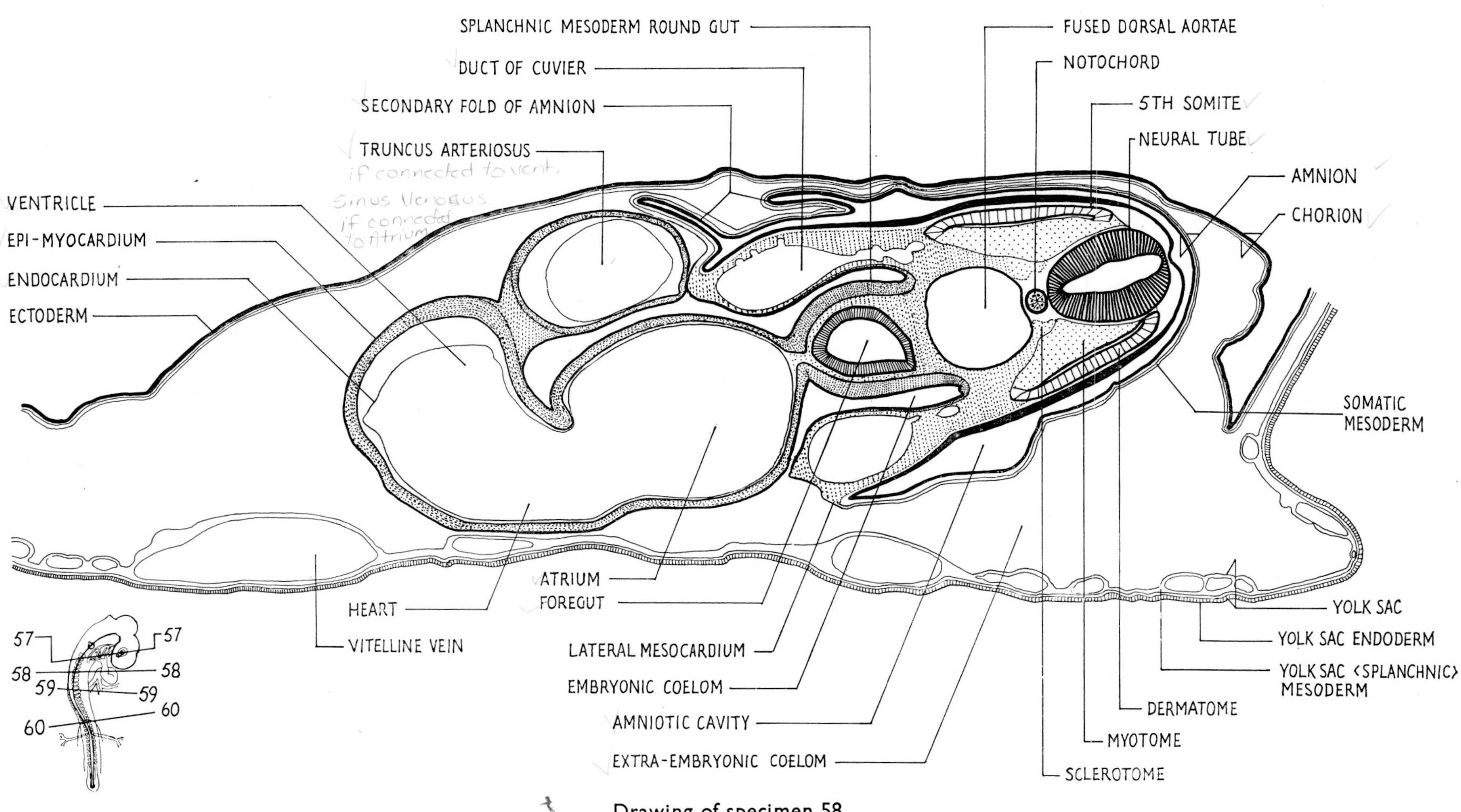

SPLANCHNIC MESODERM ROUND GUT
DUCT OF CUVIER
SECONDARY FOLD OF AMNION
TRUNCUS ARTERIOSUS
VENTRICLE
EPI-MYOCARDIUM
ENDOCARDIUM
ECTODERM
FUSED DORSAL AORTAE
NOTOCHORD
5TH SOMITE
NEURAL TUBE
AMNION
CHORION
SOMATIC MESODERM
ATRIUM
FOREGUT
HEART
VITELLINE VEIN
LATERAL MESOCARDIUM
EMBRYONIC COELOM
AMNIOTIC CAVITY
EXTRA-EMBRYONIC COELOM
YOLK SAC
YOLK SAC ENDODERM
YOLK SAC <SPLANCHNIC> MESODERM
DERMATOME
MYOTOME
SCLEROTOME
57
57
58
58
59
59
60
60

Drawing of specimen 58

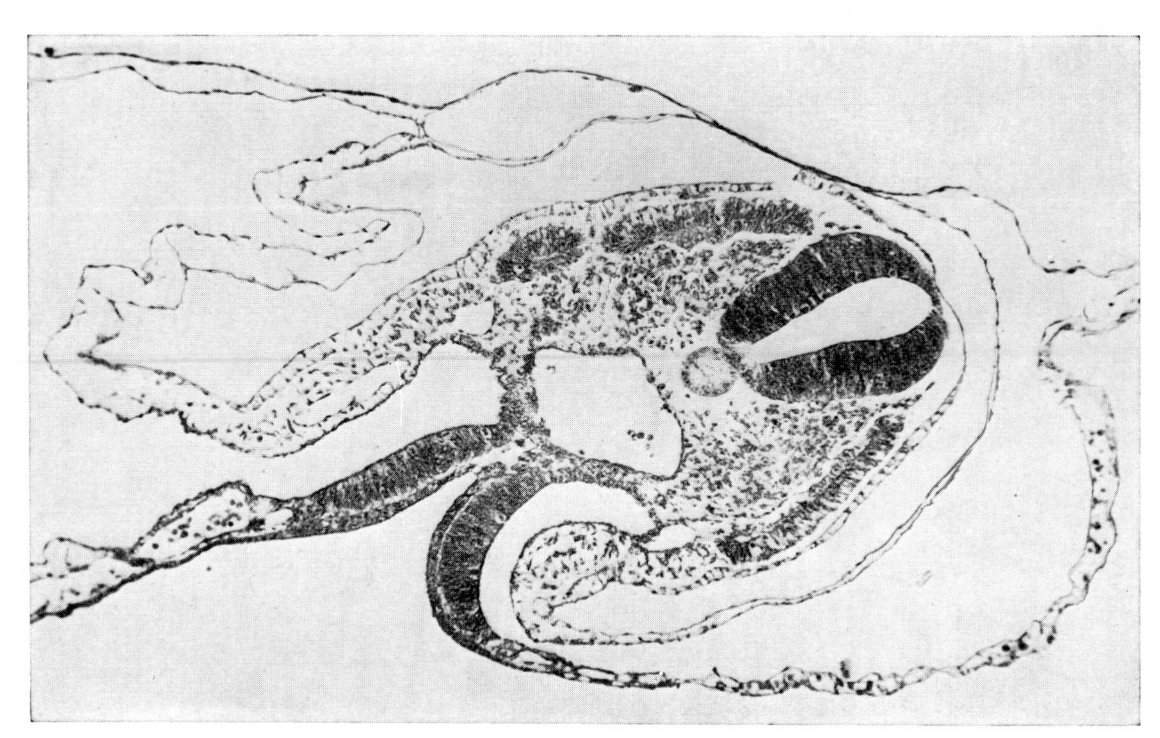

59. **Chick:** 30-somite stage, anterior trunk region, T.S. *mag. 125*×

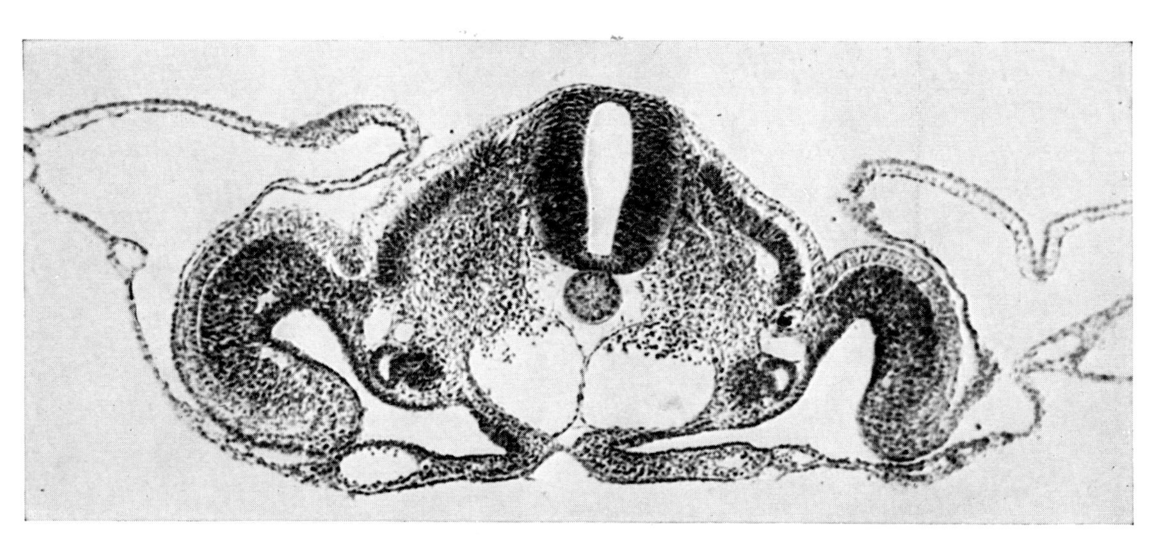

60. **Chick:** 30-somite stage, posterior trunk region, T.S. *mag. 85*×

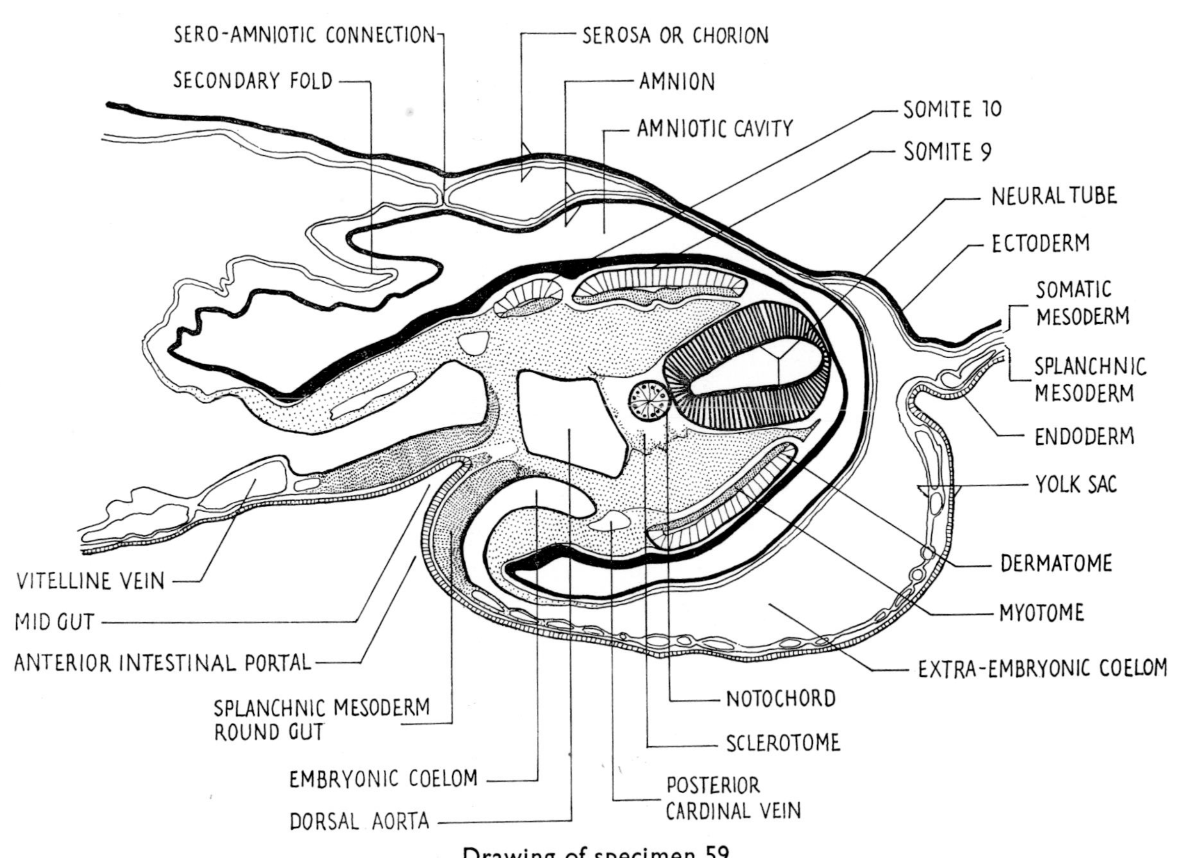
SERO-AMNIOTIC CONNECTION
SECONDARY FOLD
SEROSA OR CHORION
AMNION
AMNIOTIC CAVITY
SOMITE 10
SOMITE 9
NEURAL TUBE
ECTODERM
SOMATIC MESODERM
SPLANCHNIC MESODERM
ENDODERM
YOLK SAC
DERMATOME
MYOTOME
EXTRA-EMBRYONIC COELOM
VITELLINE VEIN
MID GUT
ANTERIOR INTESTINAL PORTAL
SPLANCHNIC MESODERM ROUND GUT
EMBRYONIC COELOM
DORSAL AORTA
NOTOCHORD
SCLEROTOME
POSTERIOR CARDINAL VEIN

Drawing of specimen 59

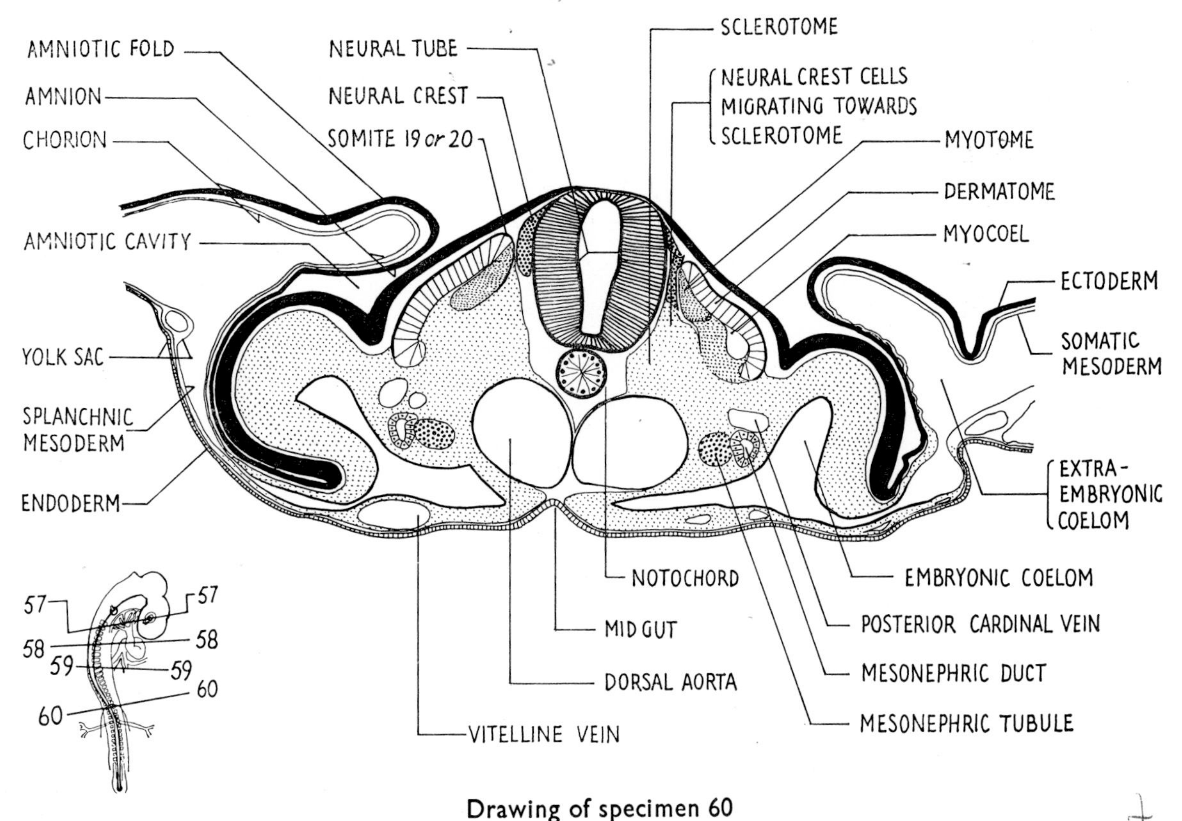
AMNIOTIC FOLD
AMNION
CHORION
AMNIOTIC CAVITY
YOLK SAC
SPLANCHNIC MESODERM
ENDODERM
NEURAL TUBE
NEURAL CREST
SOMITE 19 or 20
SCLEROTOME
NEURAL CREST CELLS MIGRATING TOWARDS SCLEROTOME
MYOTOME
DERMATOME
MYOCOEL
ECTODERM
SOMATIC MESODERM
EXTRA-EMBRYONIC COELOM
EMBRYONIC COELOM
POSTERIOR CARDINAL VEIN
MESONEPHRIC DUCT
MESONEPHRIC TUBULE
NOTOCHORD
MID GUT
DORSAL AORTA
VITELLINE VEIN
57
57
58
58
59
59
60
60

Drawing of specimen 60

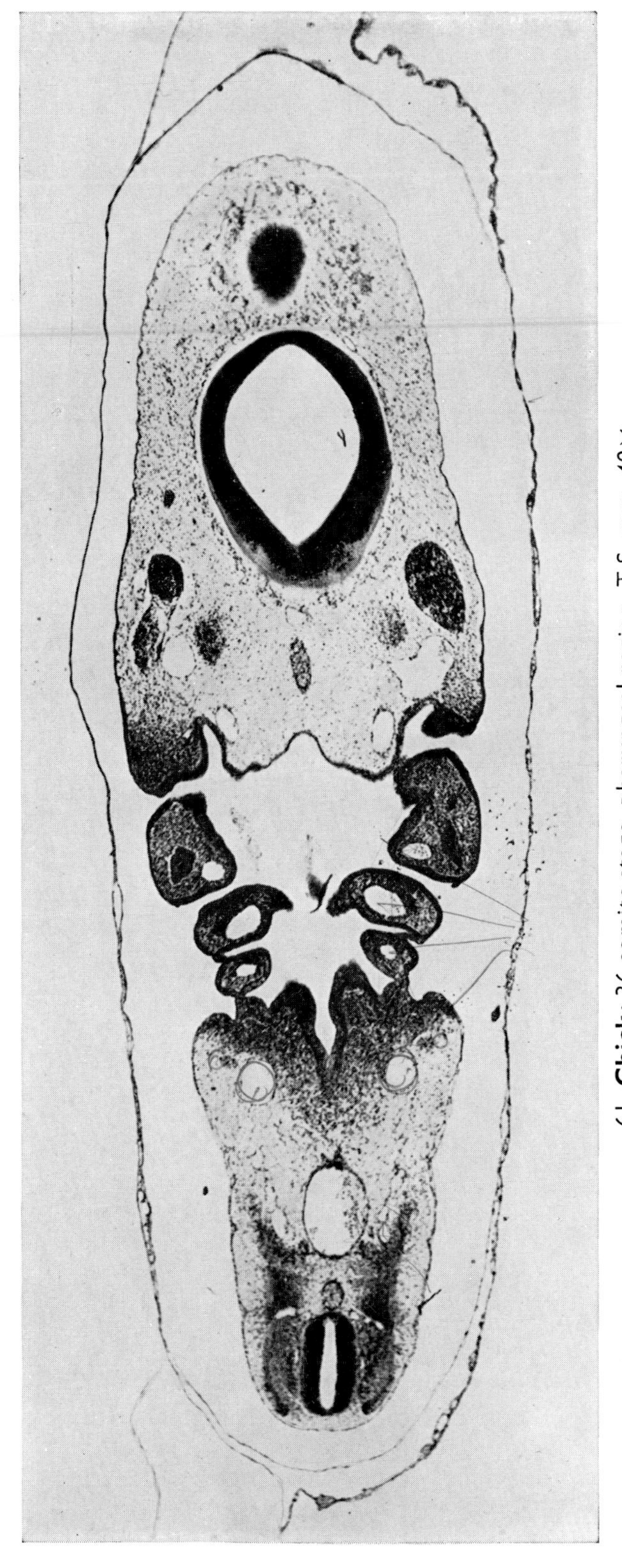

61. **Chick:** 36-somite stage, pharyngeal region, T.S. *mag.* *40*×

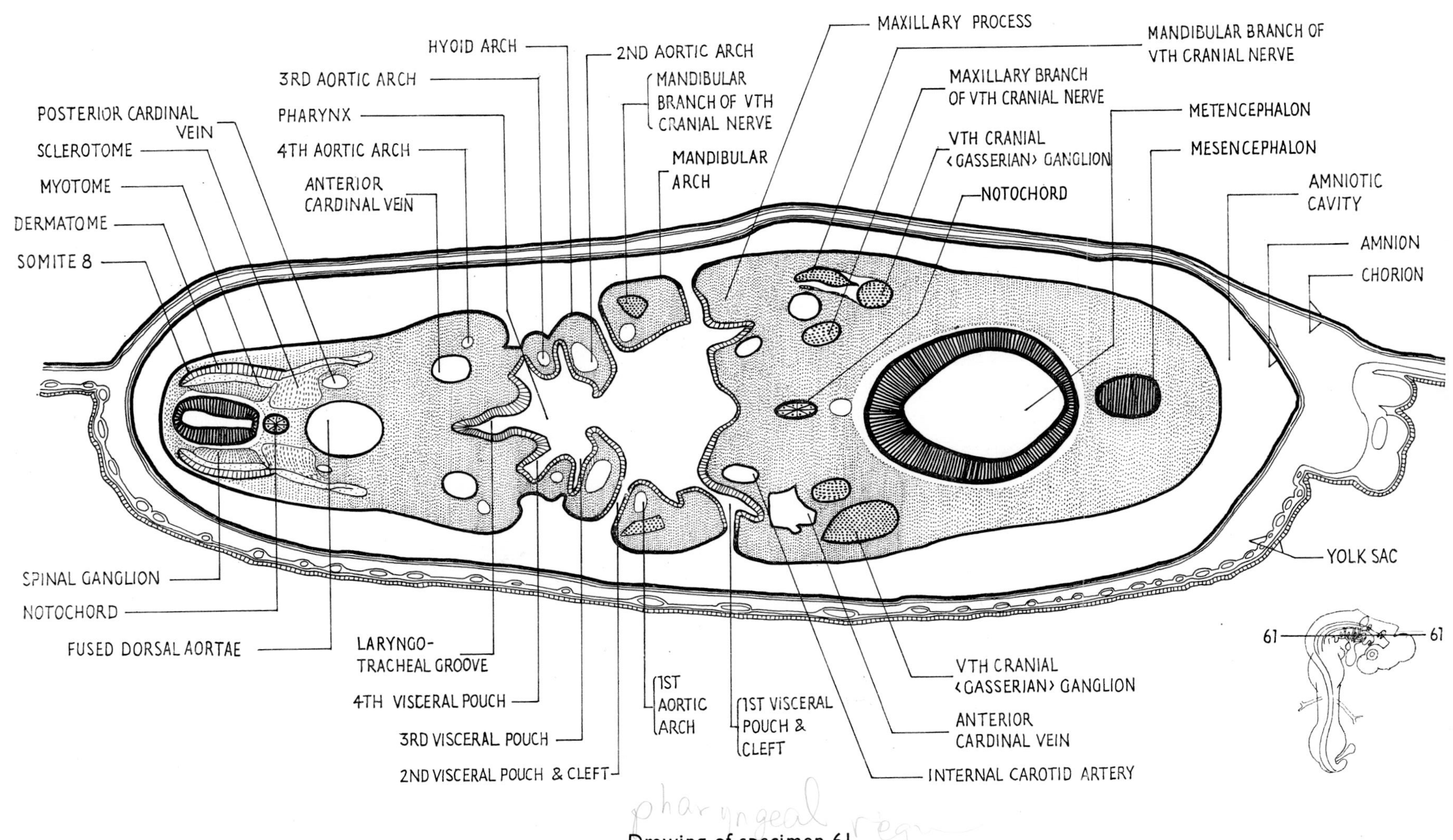

POSTERIOR CARDINAL VEIN
SCLEROTOME
MYOTOME
DERMATOME
SOMITE 8
3RD AORTIC ARCH
PHARYNX
4TH AORTIC ARCH
ANTERIOR CARDINAL VEIN
HYOID ARCH
2ND AORTIC ARCH
MANDIBULAR BRANCH OF VTH CRANIAL NERVE
MANDIBULAR ARCH
MAXILLARY PROCESS
MANDIBULAR BRANCH OF VTH CRANIAL NERVE
MAXILLARY BRANCH OF VTH CRANIAL NERVE
VTH CRANIAL <GASSERIAN> GANGLION
NOTOCHORD
METENCEPHALON
MESENCEPHALON
AMNIOTIC CAVITY
AMNION
CHORION
YOLK SAC
SPINAL GANGLION
NOTOCHORD
FUSED DORSAL AORTAE
LARYNGO-TRACHEAL GROOVE
4TH VISCERAL POUCH
3RD VISCERAL POUCH
2ND VISCERAL POUCH & CLEFT
1ST AORTIC ARCH
1ST VISCERAL POUCH & CLEFT
VTH CRANIAL <GASSERIAN> GANGLION
ANTERIOR CARDINAL VEIN
INTERNAL CAROTID ARTERY
61
61

Drawing of specimen 61

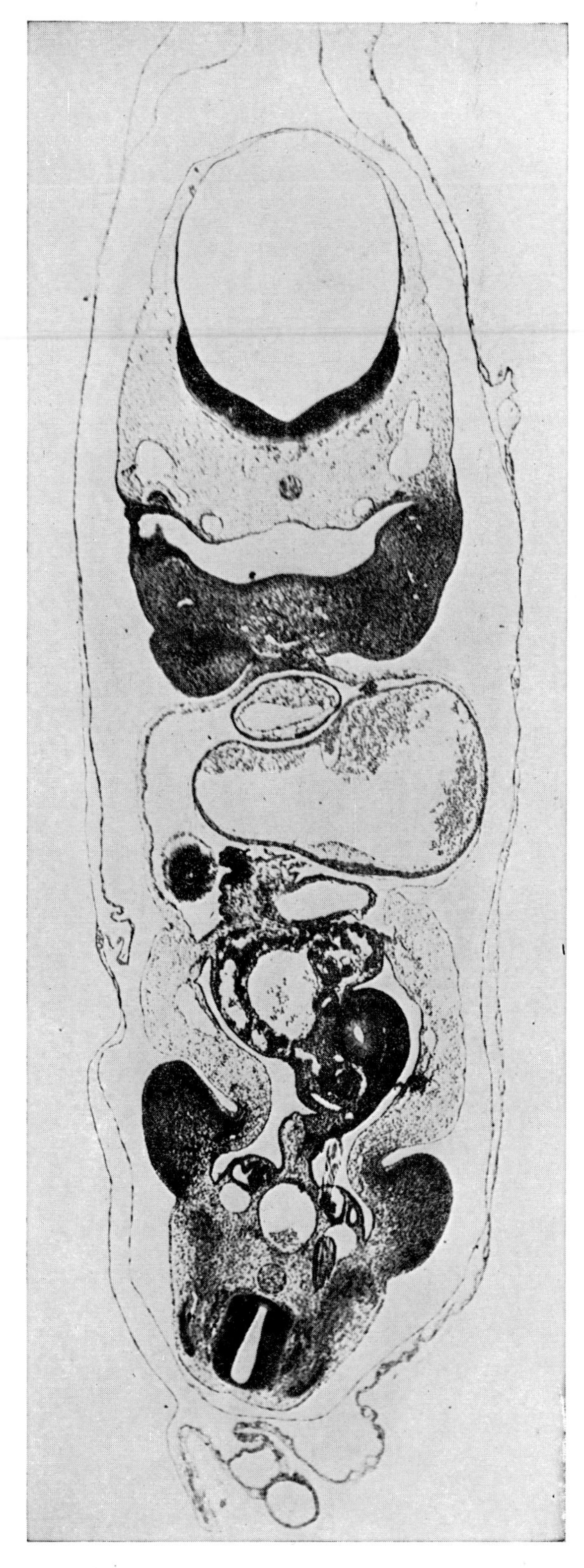

62. **Chick:** 36-somite stage, hind-brain region, T.S. *mag. 40*×

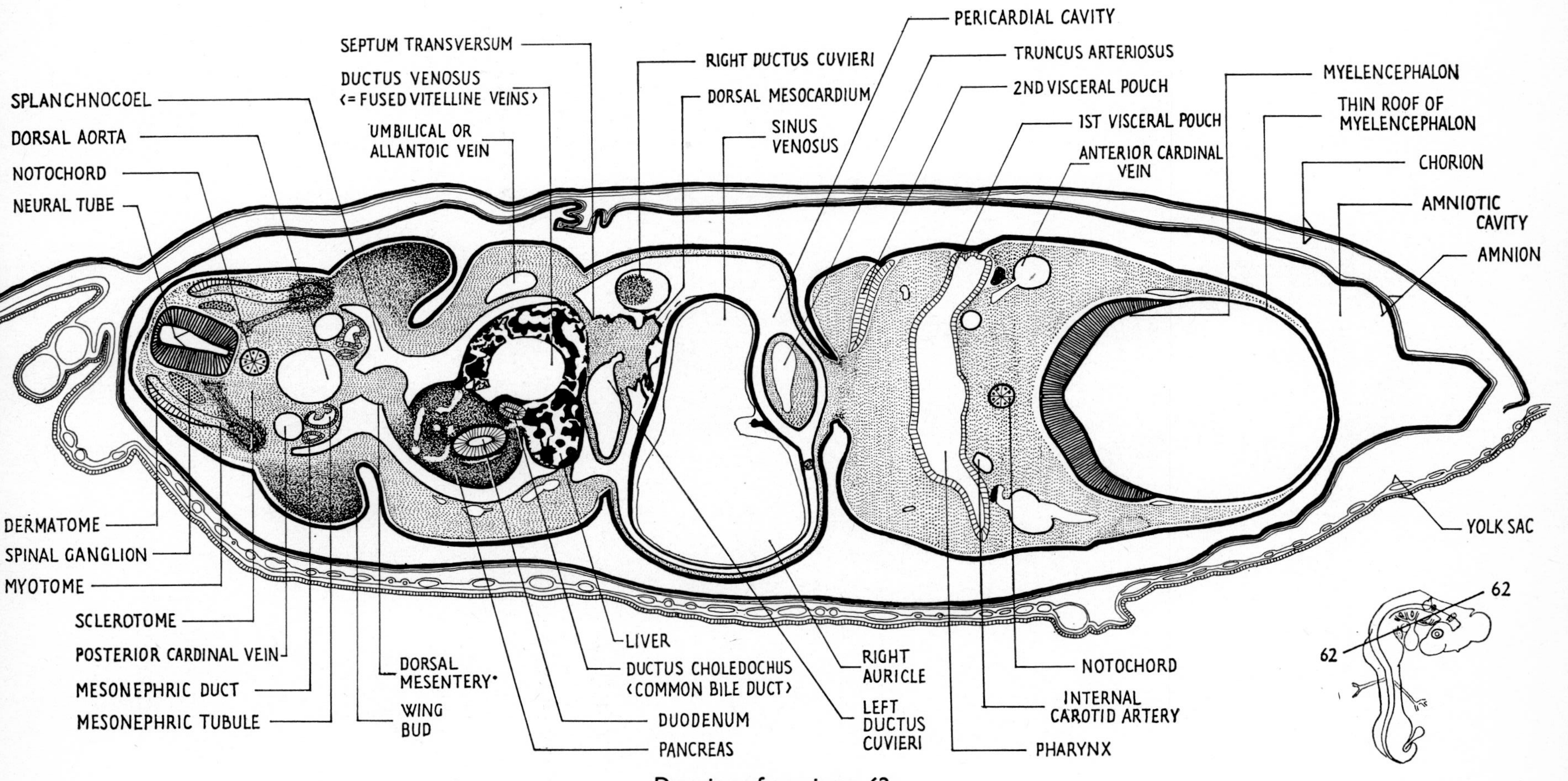

SPLANCHNOCOEL
DORSAL AORTA
NOTOCHORD
NEURAL TUBE
SEPTUM TRANSVERSUM
DUCTUS VENOSUS
(= FUSED VITELLINE VEINS)
UMBILICAL OR
ALLANTOIC VEIN
RIGHT DUCTUS CUVIERI
DORSAL MESOCARDIUM
SINUS
VENOSUS
PERICARDIAL CAVITY
TRUNCUS ARTERIOSUS
2ND VISCERAL POUCH
1ST VISCERAL POUCH
ANTERIOR CARDINAL
VEIN
MYELENCEPHALON
THIN ROOF OF
MYELENCEPHALON
CHORION
AMNIOTIC
CAVITY
AMNION
YOLK SAC
62
62
DERMATOME
SPINAL GANGLION
MYOTOME
SCLEROTOME
POSTERIOR CARDINAL VEIN
MESONEPHRIC DUCT
MESONEPHRIC TUBULE
DORSAL
MESENTERY*
WING
BUD
LIVER
DUCTUS CHOLEDOCHUS
(COMMON BILE DUCT)
DUODENUM
PANCREAS
RIGHT
AURICLE
LEFT
DUCTUS
CUVIERI
NOTOCHORD
INTERNAL
CAROTID ARTERY
PHARYNX

Drawing of specimen 62

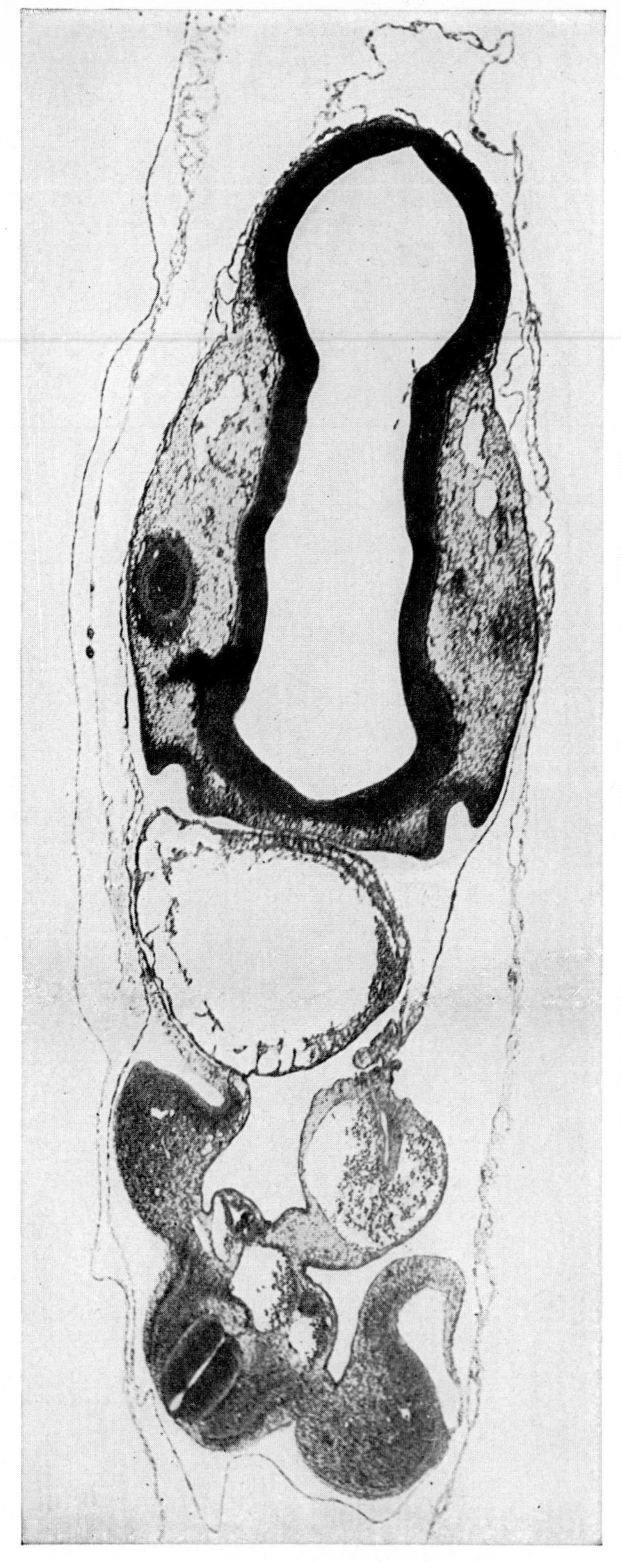

63. **Chick:** 36-somite stage, olfactory pit region, T.S. *mag.* 45×

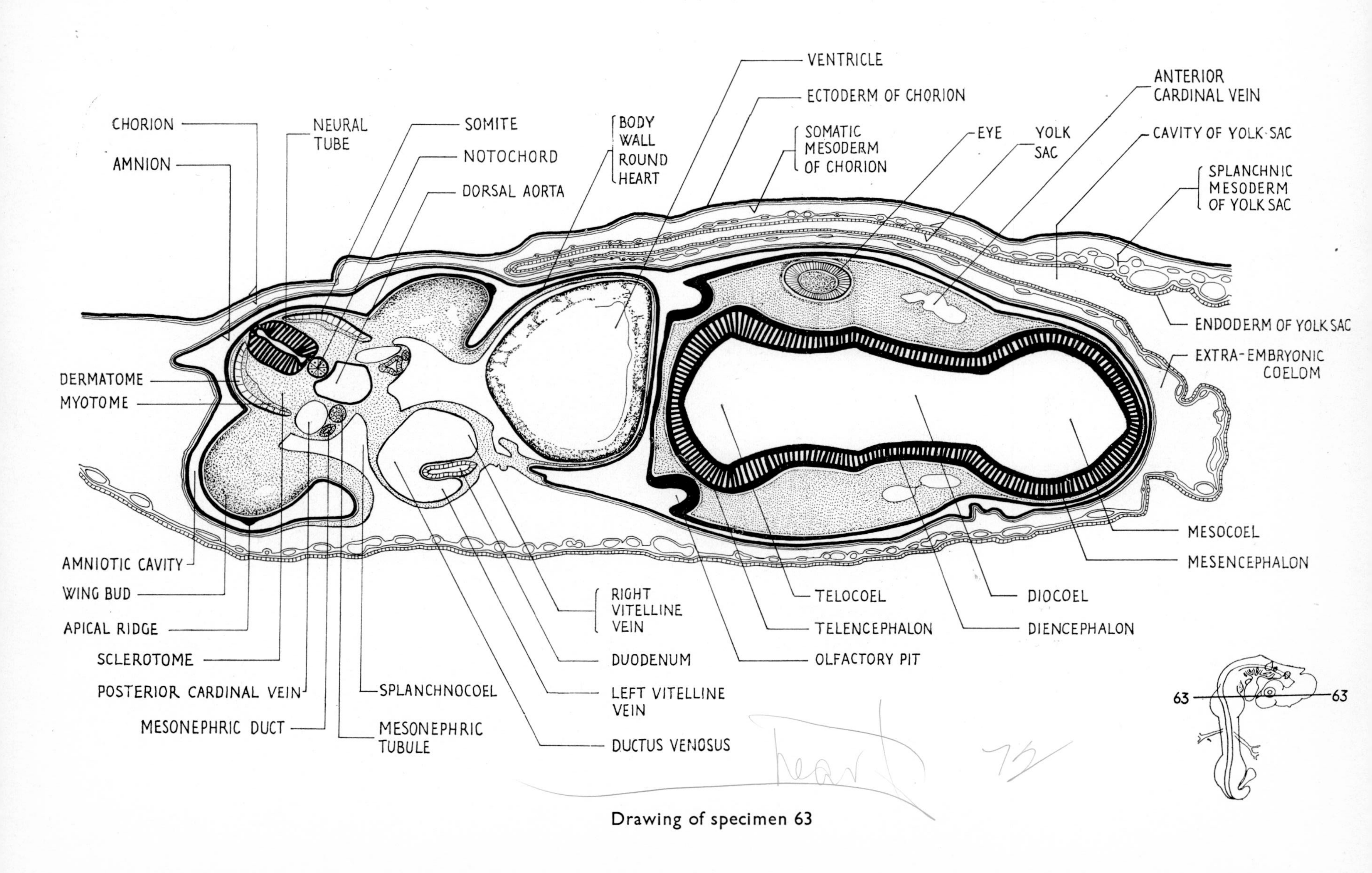
VENTRICLE
ECTODERM OF CHORION
ANTERIOR CARDINAL VEIN
CHORION
NEURAL TUBE
SOMITE
BODY WALL ROUND HEART
SOMATIC MESODERM OF CHORION
EYE
YOLK SAC
CAVITY OF YOLK-SAC
AMNION
NOTOCHORD
SPLANCHNIC MESODERM OF YOLK SAC
DORSAL AORTA
ENDODERM OF YOLK SAC
EXTRA-EMBRYONIC COELOM
DERMATOME
MYOTOME
MESOCOEL
AMNIOTIC CAVITY
MESENCEPHALON
RIGHT VITELLINE VEIN
TELOCOEL
DIOCOEL
WING BUD
APICAL RIDGE
TELENCEPHALON
DIENCEPHALON
SCLEROTOME
DUODENUM
OLFACTORY PIT
POSTERIOR CARDINAL VEIN
SPLANCHNOCOEL
LEFT VITELLINE VEIN
63
63
MESONEPHRIC DUCT
MESONEPHRIC TUBULE
DUCTUS VENOSUS

Drawing of specimen 63

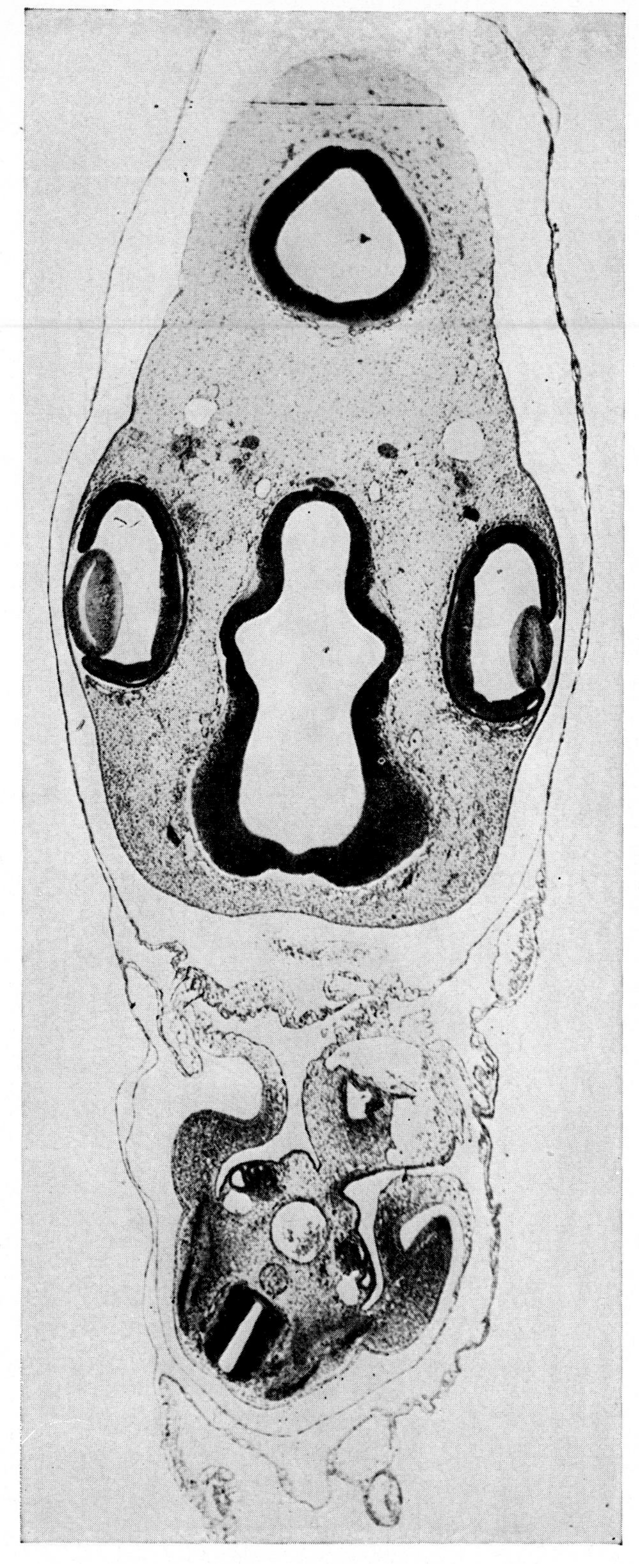

64. Chick: 36-somite stage, optic region, T.S. *mag.* 40×

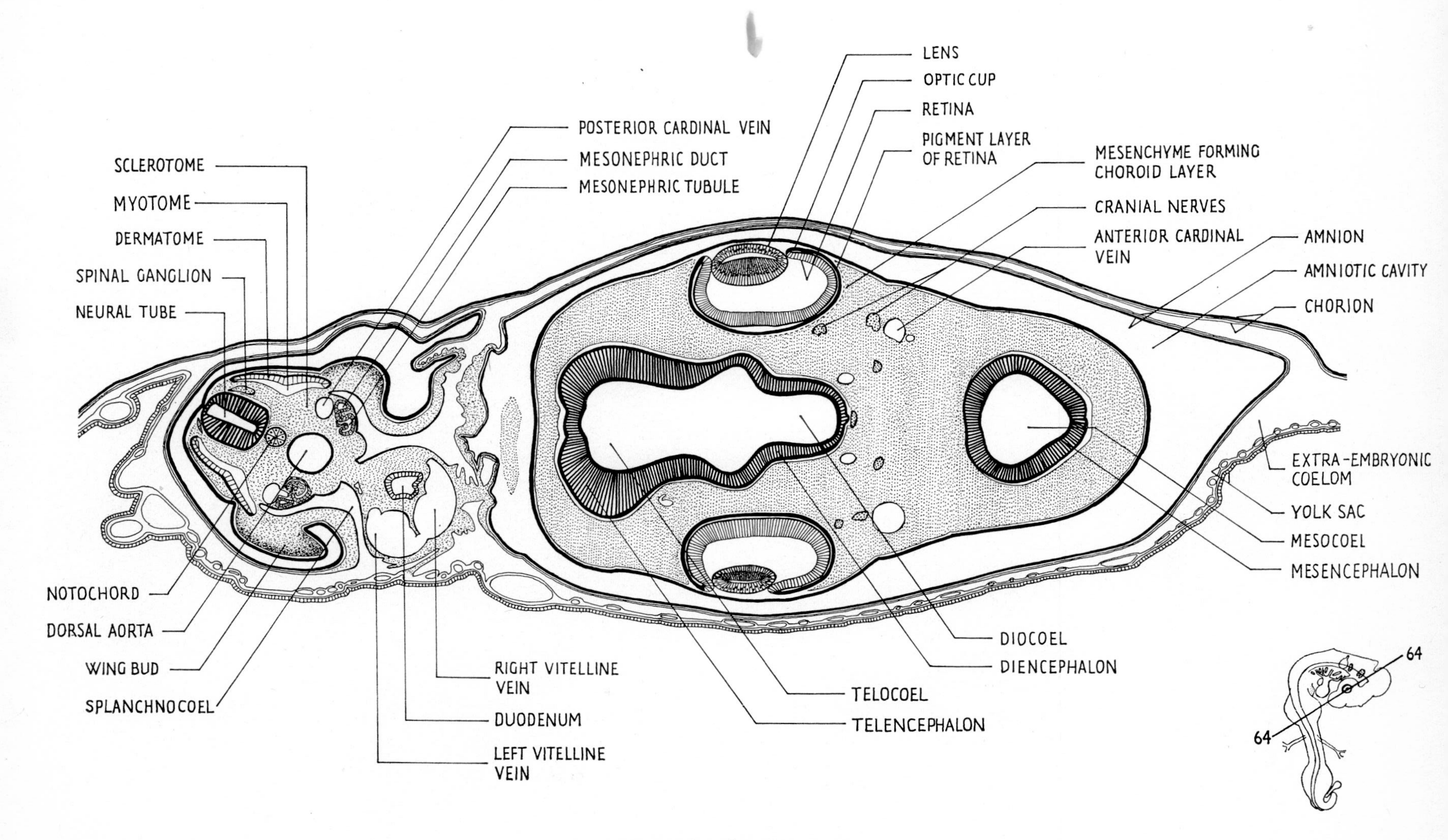
LENS
OPTIC CUP
RETINA
PIGMENT LAYER OF RETINA
MESENCHYME FORMING CHOROID LAYER
CRANIAL NERVES
ANTERIOR CARDINAL VEIN
AMNION
AMNIOTIC CAVITY
CHORION
EXTRA-EMBRYONIC COELOM
YOLK SAC
MESOCOEL
MESENCEPHALON
POSTERIOR CARDINAL VEIN
MESONEPHRIC DUCT
MESONEPHRIC TUBULE
SCLEROTOME
MYOTOME
DERMATOME
SPINAL GANGLION
NEURAL TUBE
NOTOCHORD
DORSAL AORTA
WING BUD
SPLANCHNOCOEL
RIGHT VITELLINE VEIN
DUODENUM
LEFT VITELLINE VEIN
DIOCOEL
DIENCEPHALON
TELOCOEL
TELENCEPHALON
64
64

Drawing of specimen 64

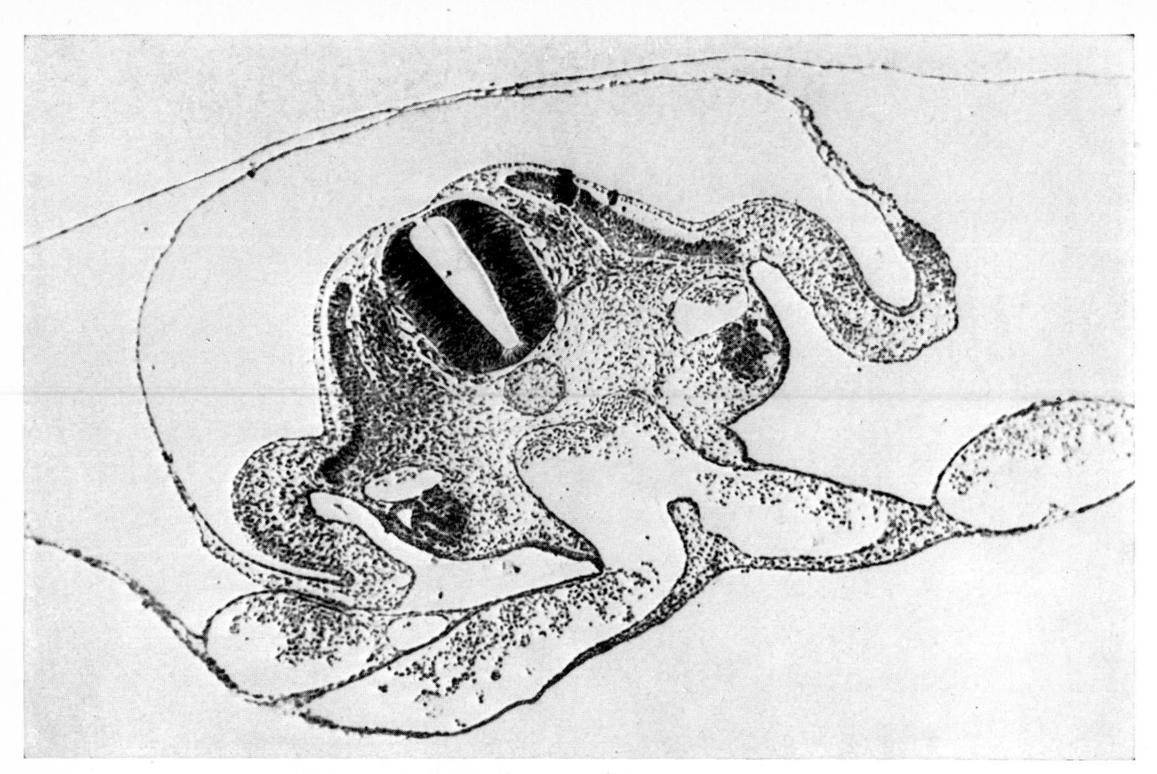

65. **Chick:** 36-somite stage, trunk region, T.S. *mag*. 75×

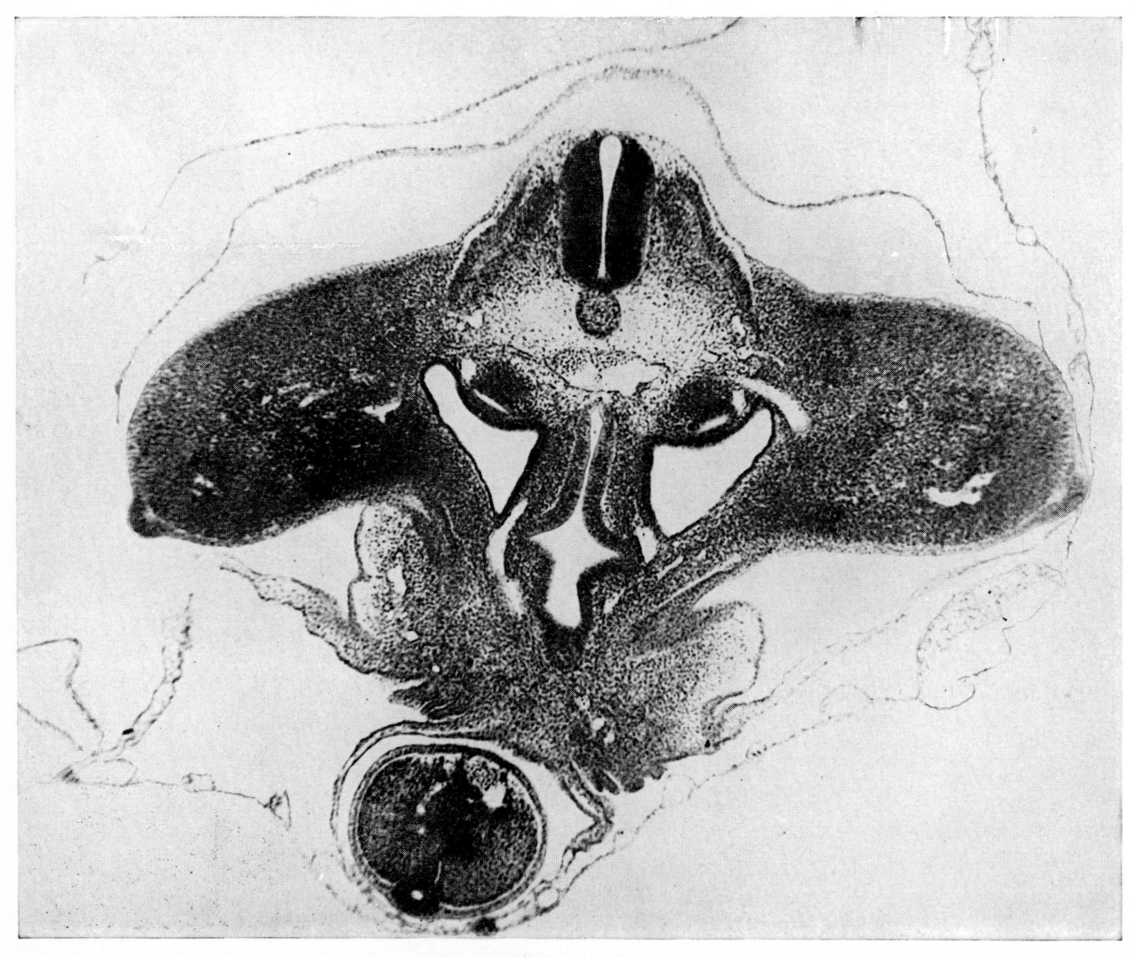

66. **Chick:** 45-somite stage, tail and hind-limb region, T.S. *mag*. 60×

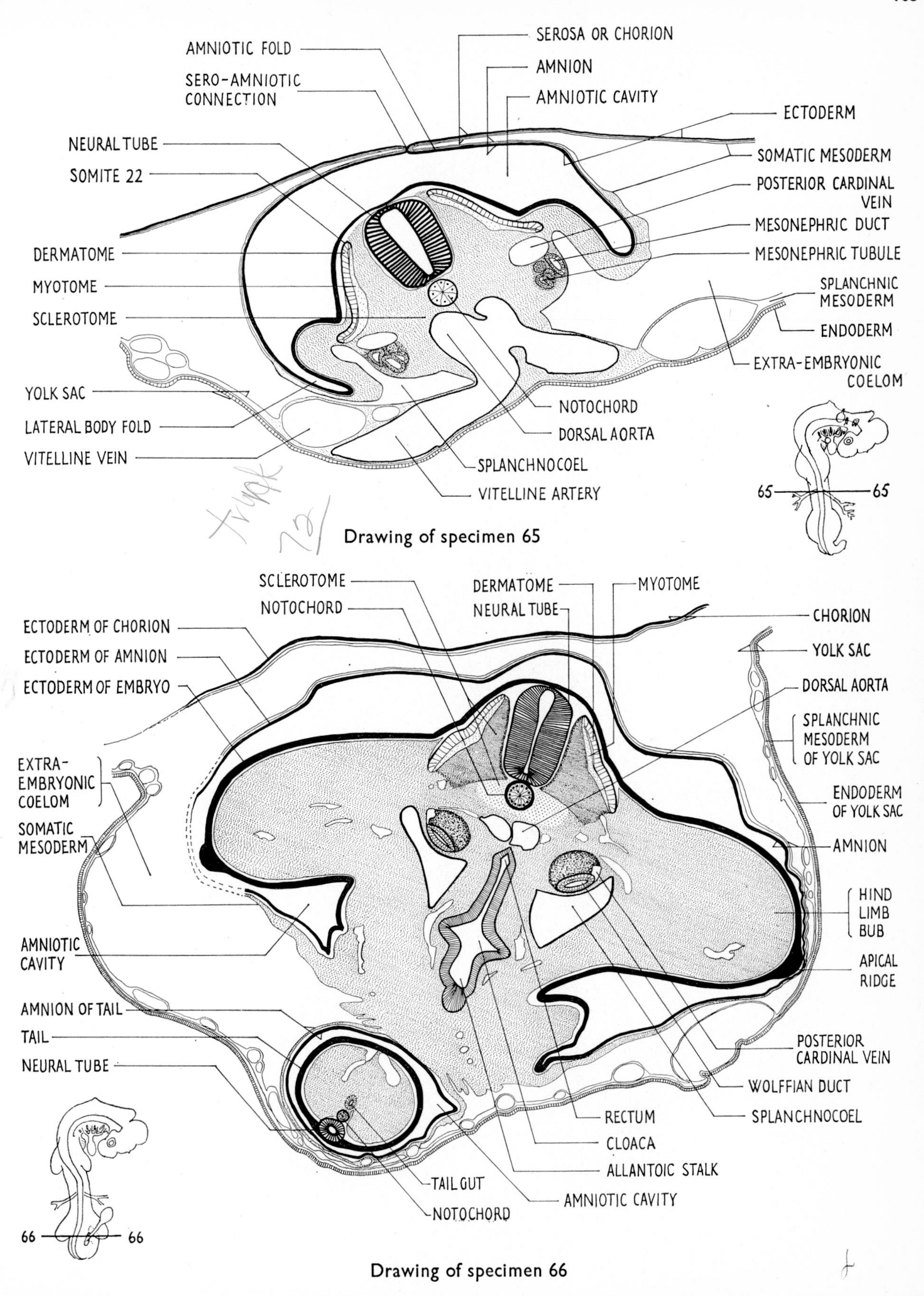
AMNIOTIC FOLD
SERO-AMNIOTIC CONNECTION
NEURAL TUBE
SOMITE 22
DERMATOME
MYOTOME
SCLEROTOME
YOLK SAC
LATERAL BODY FOLD
VITELLINE VEIN
SEROSA OR CHORION
AMNION
AMNIOTIC CAVITY
ECTODERM
SOMATIC MESODERM
POSTERIOR CARDINAL VEIN
MESONEPHRIC DUCT
MESONEPHRIC TUBULE
SPLANCHNIC MESODERM
ENDODERM
EXTRA-EMBRYONIC COELOM
NOTOCHORD
DORSAL AORTA
SPLANCHNOCOEL
VITELLINE ARTERY
65
65

Drawing of specimen 65

SCLEROTOME
NOTOCHORD
DERMATOME
NEURAL TUBE
MYOTOME
CHORION
ECTODERM OF CHORION
ECTODERM OF AMNION
ECTODERM OF EMBRYO
YOLK SAC
DORSAL AORTA
SPLANCHNIC MESODERM OF YOLK SAC
EXTRA-EMBRYONIC COELOM
SOMATIC MESODERM
ENDODERM OF YOLK SAC
AMNION
HIND LIMB BUB
AMNIOTIC CAVITY
APICAL RIDGE
AMNION OF TAIL
TAIL
NEURAL TUBE
POSTERIOR CARDINAL VEIN
WOLFFIAN DUCT
SPLANCHNOCOEL
RECTUM
CLOACA
ALLANTOIC STALK
TAIL GUT
AMNIOTIC CAVITY
NOTOCHORD
66
66

Drawing of specimen 66

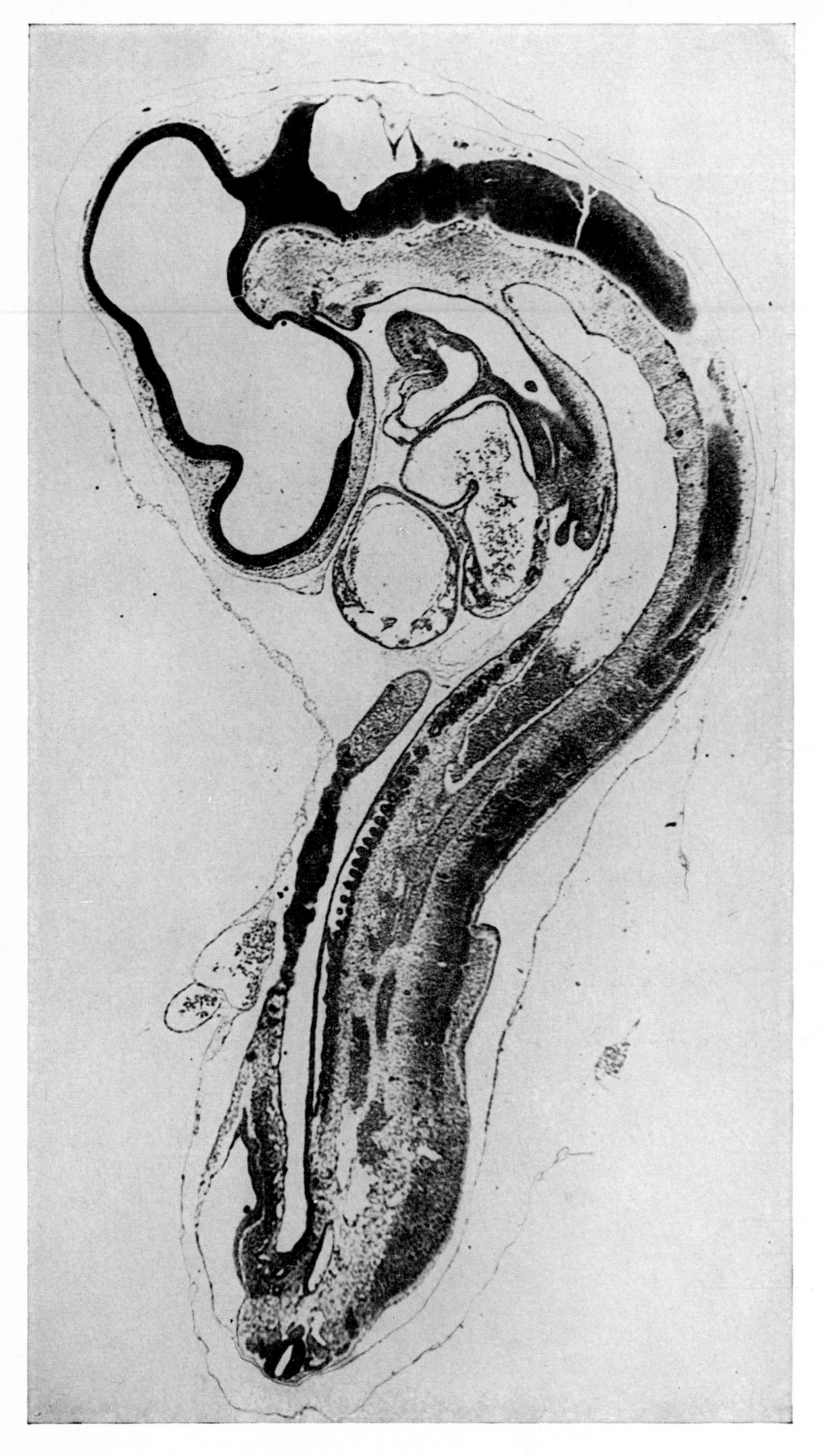

67. **Chick:**
36-somite stage,
H.L.S.
mag. 25×

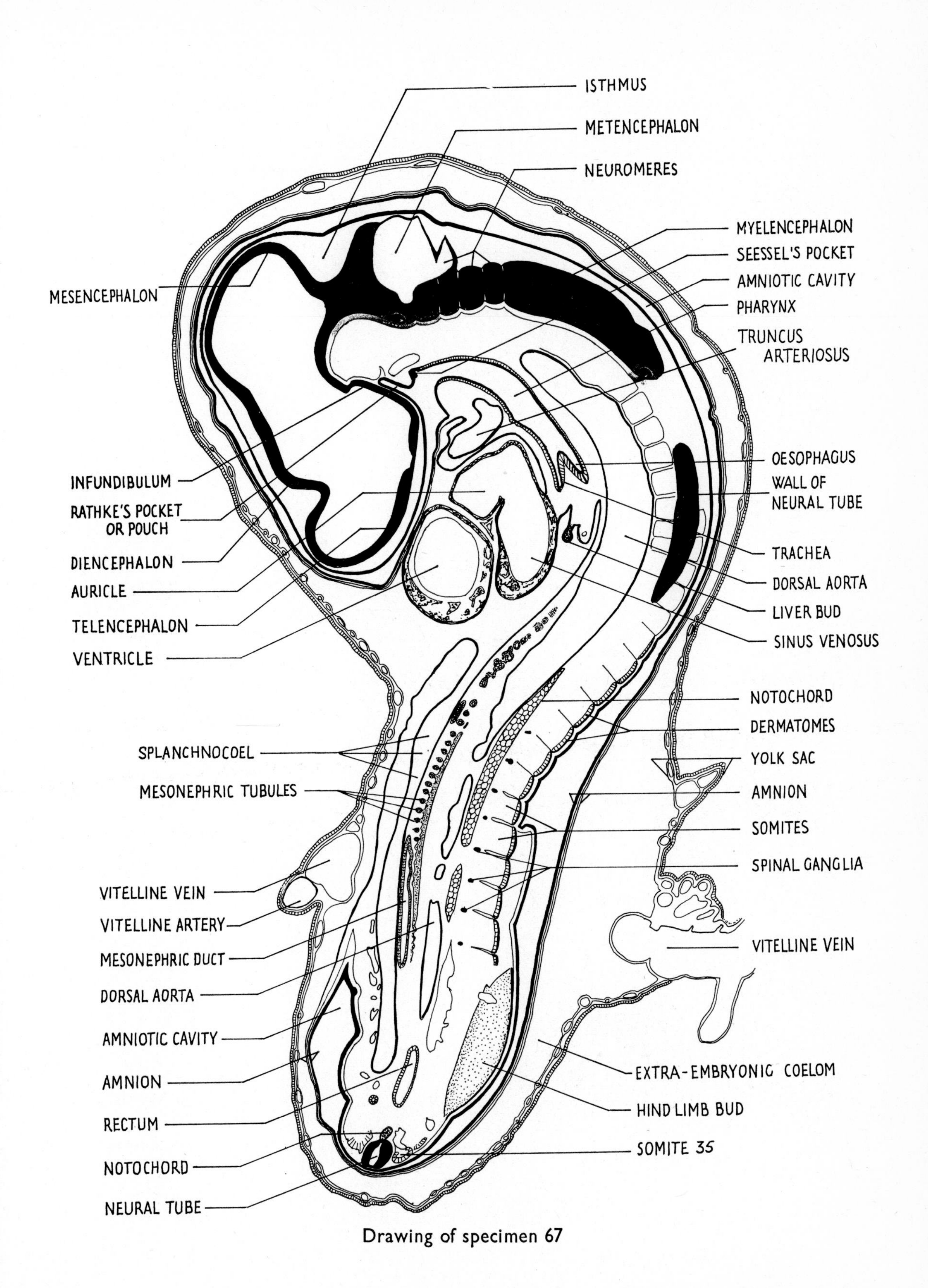
ISTHMUS
METENCEPHALON
NEUROMERES
MYELENCEPHALON
SEESSEL'S POCKET
AMNIOTIC CAVITY
PHARYNX
TRUNCUS ARTERIOSUS
MESENCEPHALON
OESOPHAGUS
WALL OF NEURAL TUBE
INFUNDIBULUM
RATHKE'S POCKET OR POUCH
TRACHEA
DIENCEPHALON
DORSAL AORTA
AURICLE
LIVER BUD
TELENCEPHALON
SINUS VENOSUS
VENTRICLE
NOTOCHORD
DERMATOMES
SPLANCHNOCOEL
YOLK SAC
MESONEPHRIC TUBULES
AMNION
SOMITES
SPINAL GANGLIA
VITELLINE VEIN
VITELLINE ARTERY
MESONEPHRIC DUCT
VITELLINE VEIN
DORSAL AORTA
AMNIOTIC CAVITY
EXTRA-EMBRYONIC COELOM
AMNION
HIND LIMB BUD
RECTUM
SOMITE 35
NOTOCHORD
NEURAL TUBE

Drawing of specimen 67

A REFERENCE TABLE OI

Number of somites	Stage*	Incubation time in hours according to:— Duval	Huettner	Patten	Lillie	Primitive streak	Nervous system	Mesoderm, somites and kidney	Vascular system	Anterior intestinal portal
0	*4*	*20*	*17–18*	*18*	*18–19*	*Maximal length, 2.2 mm., i.e., 0.7 of area pellucida. Groove, pit and node present.*		*Shield shaped sheet of mesoderm spreads out laterally from the primitive streak.*		
0	*5 & 6*	*21*	*19*	*20*	*19–22*	*Begins to decrease in length, 1.9 mm. Notochord grows forward from node.*	*Neural plate and neural folds visible.*	*Lateral horns of mesoderm grow forward. The first somite may appear simultaneously with the formation of the head fold (stage 7).*	*Mesenchyme cells form isolated blood islands in extra-embryonic mesoderm.*	*First seen to be present.*
3	*8–*	*22*	*23*	*23*	*25–28*	*Reduced to a length of 1.5 mm.*	*Neural folds meet in brain region but do not fuse.*	*Lateral horns grow round the mesodermless proamnion. Segmented somites joined to lateral plate mesoderm by intermediate mesoderm (nephrotome). A cavity, the myocoel, appears in somites.*	*The blood islands begin to unite and the first blood corpuscles are produced within the resulting tubes.*	*Moves back as the foregut elongates.*
5	*8+*	*23–25*	*25*	*25–26*	*27–30*	*1.2 mm. long.*	*Fusion of folds begins in brain region; further back neural folds meet but they splay out over the somites.*	*The cells of the somites become radially arranged about the myocoel cavities; cavity reduced by a central core of cells. Lateral horns meet anteriorly.*	*The embryo becomes linked to the blood island system by vitelline veins. Paired primordia of the heart develop together with ventral and dorsal aortae.*	*Lies posterior to the heart primordia.*
10	*10*	*29–30*	*30*	*30–31*	*33–38*	*0.6 mm. long.*	*Except for anterior neuropore, fusion of folds is completed in the brain region. Three primary brain vesicles visible.*	*The intermediate mesoderm begins to separate off dorsally. The pronephric tubules develop from this material between somites six and ten. The first somite begins to disappear.*	*The heart primordia fuse to form a tubular heart which bends slightly to the right of the embryo. Faint and sporadic pulsation of the heart occurs.*	*May reach the first somite.*
13	*11*	*33–34*	*33*	*33–34*	*40–45*	*0.4 mm. long.*	*Five brain vesicles can be seen. Anterior neuropore closes. The neural folds fuse beyond the thirteenth somite.*	*The dorso-lateral buds differentiate into pronephric tubules and the pronephric duct forms by fusion of material from the tubules. First signs of Wolffian duct.*	*The heart becomes distinctly displaced to the right. The rate and amplitude of the heart beats increase. A network of blood vessels established in area vasculosa.*	*Reaches the second somite.*
17	*12+*	*37–41*	*37*	*38–40*	*46–50*	*0.2 mm. long.*	*Fore brain at an angle to hind brain due to flexure. A shallow infundibulum is present.*	*Connection between somites and nephrotomes is lost. The mesonephros develops along with pronephros below the somites. Wolffian duct extends from tenth to fifteenth somite. Differentiation begins in anterior somites.*	*The heart is beating efficiently by this stage and blood circulates. The heart is S-shaped. The first aortic arch begins to develop. The dorsal aortae fuse between somites three and four. The vitelline artery can be seen between somites sixteen and seventeen.*	*Reaches the third somite.*
21	*14+*	*43–46*	*43*	*44–48*	*48–52*	*No longer distinguishable: contributes material to tail bud.*	*Fore brain at right angles to hind brain. Fore brain enlarges in telencephalon region.*	*Pronephros begins to disappear anterior to the eleventh somite. In the anterior somites a distinct dermatome can be seen and cells migrate from the somites and neural crests to form the sclerotomes round the notochord.*	*The atrium begins to divide into right and left auricles. The first aortic arch is established and the second begins to form. Fusion of dorsal aortae may reach somite eight. The vitelline artery is distinct between somites 17–19.*	*Reaches the fourth somite.*
24	*15*	*44–46*	*48*	*48–50*	*50–55*		*The telencephalon becomes distinct from the diencephalon. Rathke's pocket grows under the infundibulum.*	*The posterior somites remain undifferentiated; anteriorly somites differentiate into dermatome, myotome and sclerotome. There are eleven pairs of mesonephric tubules between somite five and sixteen.*	*Besides the two auricles heart has distinct ventricle and conus arteriosus. Two aortic arches present. Dorsal aortae fuse as far back as somite twelve. The vitelline arteries lie between somites eighteen and twenty.*	*Is in the region of somites five to six.*
27	*16*	*48*	*50–52*	*50–55*	*51–56*		*Telencephalon and diencephalon become separated by the velum transversum. A distinct isthmus can be seen between the mesencephalon and metencephalon.*	*Differentiation into dermatome, myotome and sclerotome reaches somite twenty. Wolffian duct and mesonephric tubules seen in trunk sections.*	*The third aortic arch appears. The dorsal aortae fuse between somites four and fourteen. The vitelline artery lies between somites 19 and 21. Vitelline veins join to form ductus venosus which opens into sinus venosus.*	*Lies between somites seven and ten.*
30	*17*	*52*	*58–60*	*55–60*	*52–64*		*The isthmus deepens. Paired telencephalic vesicles develop. Roof of hind brain becomes very thin in myelencephalon region. Brain bent double by now.*	*Differentiation reaches the twenty-fifth somite. Wolffian duct grows back towards cloaca. Glomeruli can be seen in mesonephric tubules.*	*There are three complete aortic arches and the fourth begins to develop. The first pair of aortic arches may begin to atrophy at this stage. Dorsal aortae fused up to somite 16. Vitelline artery between somites 20 and 22.*	*Has moved back to lie between somites ten and twelve.*
36	*18+*	*68–72*	*72*	*72*	*72*		*The cerebral hemispheres develop from the telencephalic vesicles. The infundibulum joins with Rathke's pocket to form the pituitary.*	*Differentiation reaches the thirtieth somite. Wolffian duct reaches cloaca but may not fuse with it until later.*	*The first pair of aortic arches continue to atrophy as the fourth pair develop. Dorsal aortae fused as far back as somites 17–20. Vitelline artery is in region of somites 21–22.*	*Between somites thirteen and fourteen.*

* Hamilton & Hamburger